D1320351

RELIGIOUS LIFE AND ENGLISH CULTURE IN THE REFORMATION

Religious Life and English Culture in the Reformation

Marjo Kaartinen

Senior Research Fellow
University of Turku
Finland

First published 2002 by
PALGRAVE
Houndmills, Basingstoke, Hampshire RG21 6XS and
175 Fifth Avenue, New York, N. Y. 10010
Companies and representatives throughout the world

PALGRAVE is the new global academic imprint of
St. Martin's Press LLC Scholarly and Reference Division and
Palgrave Publishers Ltd (formerly Macmillan Press Ltd).

ISBN 0–333–96924–3

This book is printed on paper suitable for recycling and made from fully managed and sustained forest sources.

A catalogue record for this book is available from the British Library.

Library of Congress Cataloging-in Publication Data
Kaartinen, Marjo, 1964–
    Religious life and English culture in the Reformation / Marjo Kaartinen
        p. cm.
    Includes bibliographical references and index.
    ISBN 0–333–96924–3
    1. Monasticism and religious orders—England. 2. Reformation.
    3. Monasticism and religious orders—History—Middle Ages, 600–1500.
    4. England—Church history—1066–1485. I. Title.

BX2592 .K37 2002
271'.00942'09031—dc21   2001058068

10   9   8   7   6   5   4   3   2   1
11   10   09   08   07   06   05   04   03   02

Printed and bound in Great Britain by Antony Rowe Ltd, Chippenham, Wiltshire

To my son

Contents

Acknowledgements

I owe a great debt to a number of people and institutions for their encouragement and support. This book would not have been possible without my inspiring academic partnership with Anu Korhonen. Keijo Virtanen, Veikko Litzen, Kari Immonen, Natalie Zemon Davis, Sally M. Miller, Keith Battarbee, and the anonymous readers of my manuscript all enriched my vision in the process of writing this book. It goes without saying that I am deeply grateful to all my colleagues and students at the Department of Cultural History of the University of Turku.

Several institutes and their staff deserve my deepest gratitude. I wish to thank the British Library, the University Libraries at Turku, Åbo Akademi, Helsinki, Groningen and the University of London, as well as the Warburg Institute Library, and the 1995 staff of the Finnish Institute in London. I am also grateful for the help I have received at the School of Eastern European and Slavonic Studies and the Historical Institute of the University of London, the Warburg Institute, and the Department of English of the University of Helsinki.

For generous financial support I wish to thank the Finnish Cultural Foundation, the Oskar Öflund Foundation, the Alfred Kordelin Foundation, the Emil Aaltonen Foundation, the Ella and Georg Ehrnrooth Foundation, and especially the Academy of Finland. I also want to thank the Finnish National Gallery for their kind permission to use Turku-born Lauréus's painting on the jacket.

Without my family and friends, however, nothing would have been possible. A warm thank you is due to my encouraging mother, and no words are enough to express my gratitude to my partner Sanna-Kaisa Tanskanen. I wish to dedicate this book to our son, Lauri Tanskanen.

Marjo Kaartinen
Pasadena, California, December 2001

Introduction

But hither, to religion, thou shuldest harken,
hither thou shuldest loke,
hyther thou shuldeste enclyne thy mynd:
endeuoyre thy selfe with all thy myght,
imploy hereon sharpely thy wytte.

Erasmus: *De contemptu mundi*, f. 15v

Flee flee (wylte thou that I shal speke it a thousand tymes?)
flee and eschewe the estate of relygyous men,
which is so chosen, sought and inuented by the madnes and
vanytie
of mannes counsayll, without scripture, without the byddyng,
without the comaundemente and worde of god.

The Images of a verye Chrysten bysshop, sig. H5

The religious life was heaven and hell; religious houses could be depicted as ideal homes or centers of madness. For the likes of St Augustine, monasteries were the sweet smell to God, or they could be compared to a pleasant and precious wine [1] Regardless of how it was perceived, monasticism was a focal point in late medieval religious culture. Moreover, with religious houses, villages and cities were born and existing ones reformed, and religious life structured the everyday life of their inhabitants. At some point in their lives, everyone had contact with religious houses, and and monks, nuns, or friars. This book is an exploration of attitudes toward the religious life during the reign of Henry VIII (1509–47), an attempt to capture the mentality of the people who lived in those tumultuous times. Below, I will attempt to interpret, give meaning to and find significance in the thoughts of the people of the past.[2]

Cultural history always perceives human life as a social and as a cultural construction; an understanding of the people of the past is essential if we want to understand their culture.[3]

Caroline Walker Bynum aptly notes that through wonder or amazement, historians can find 'understanding, a significance, that is always just a little beyond our theories and our fears.'[4] As we can see, she is looking for significance and meaning as well. I believe that these can be found by examining attitudes and opinions; therefore, the present study is an investigation of attitudes of English people toward religious life, and toward religious people: monks, nuns, and friars. I will ask: How was religious life perceived? What meanings did the religious way of life have in people's minds? What was thought of the virtues of the religious? Did religious pursue these virtues correctly? Answering these questions, however, is not sufficient to gain a full understanding of early sixteenth-century religious culture. Therefore, the question 'Why?' is an important one. In order to be able to explain the images of the religious, I will analyze them by exploring the key concepts of the religious life and constructing a wide context for the study. The context needed here is not only late medieval religious culture, religious mentalities, and the code of ethics which the Church and the King promoted; it is also the context of the distribution of power, both secular and sacred, because it is important to understand the political background in which people lived. In this way the context unavoidably becomes itself the focus of the study.[5]

In tracing the culture of the past the main tool in this study is the analysis of the system of meanings and the culture of thought. By studying different types of texts I am searching for the conceptions of the English people of religion and the practices of religious life.[6] The texts, regardless of their role, function, or genre in the past, definitely are parts of the world in which they were created; and because they are a part of that world, they speak of that world. My sources, at which we will take a closer look later, are admittedly a varied collection of texts of different genres, conventional, traditional, and in many way ambiguous. At first glance, they present English people's attitudes toward religious life in a severely polarized light: some people considered religious houses important; others did not. It is my purpose, however, to bypass this contrastive surface level. This will enable us to see what people in opposing camps have in common: what was in common is the core of thinking in the past. What was in common was their most enduring culture, perhaps even their mentality. What remained as differing opinions is the material that molded the world into a new one slowly, over a long period of time. It is the historian's duty to recognize these two elements – in this study the enduring culture surpasses the processes of change in importance. This is a way to see why the Reformation was as successful as it was and why the fate of religious was such as it was.

Meanings constructed the past culture; therefore, texts are keys to understanding the past. Since my focus is on thinking, it follows that this study does not deal with questions such as 'What *were* the monks like?' or 'What *took place* in religious houses in 1535?' Instead, I am interested in what was *thought* to be happening, and in what the monks, nuns, and friars, their lives and their ideals, were *thought* to be like.

I would like to stress that, regardless of the emphasis on thinking in my study, we should not ignore the fact that early modern religion and religiosity were highly pragmatic. Religion, including contemplative religion, was based on performance, religious services, rites, and prayers. When this is kept in mind, it is easier to understand the medieval usage of the word 'religious' as a noun. A religious was a person who indulged in these performances.

*

The goal of this work is to find the significance of the words written (and sometimes spoken) in the past, which at first glance may say nothing of significance to us. Here what have been called the historian's clues[7] come to our aid. They disclose the past and make it intelligible. Clues can be found in all sources, in all choices of words. *Signs* are good clues: they are essential keys to culture, since people sign them and make their language understandable. For example, a sandal is not only foot-wear to protect the foot, it can also indicate a friar, or be a sign of mortification of the flesh. Signs reveal the systems that organized people's thoughts in the past.[8]

This book is constructed around certain specific 'signs,' which are the monastic vows and virtues: obedience, poverty, chastity, and stability.[9] I believe that these four aspects of religious life formed the core of all religious strategies in the early sixteenth century. The religious vows had often been under dispute, of course. For example, many humanists denied their importance, and evangelicals insisted that monastic vows had been devised by man and were thus idolatrous and blasphemous.[10] In Part I, the virtue of *obedience* and the vice of disobedience are given the main role. This chapter discusses the role of pride in the destruction of proper obedience, and explains why renouncing one's will was considered a perfect goal in life. The sovereign, whether king, pope, or abbot,[11] was expected to be revered with due obedience. Chapter 2 consequently discusses the role that obedience had to play when it came to choosing between king and pope, and the allegiance of the religious to the secular power. Part I ends with a consideration of the role of obedience in religious houses in chapter 3. Were the monks and nuns represented as disobedient? What purposes did these views serve? Did politics have a role to play in attitudes toward religious?

Poverty was a Christian virtue, and a goal that caused unending trouble for paupers and rich men alike. Part II probes the ambiguous relationship of religious to property and poverty. First, in chapter 4, problems and attitudes relating to ownership are scrutinized: How could a rich man be poor? How could a pauper own land? What did people say about monastic landownership and its consequences for their lives? Then, chapter 5 discusses the representations of monastic property: What did crystal or heated rooms signal to people? Were they signs of decadence or of devotion to God? How did the monastics themselves conceive their rich abbeys?

Part III analyzes various aspects of *chastity*, which for many was the most important facet of monasticism. I concentrate on the questions of abstinence and mortification, and their different practical forms. Representations of clothing and nourishment will be analyzed in the context of the virtue of chastity. The discussion ends with an exploration of sexuality in religious life: What was the role of virginity in people's minds? How could a monk or a nun remain chaste? Were religious people represented as whores and sodomites? If so, what did this mean in early sixteenth-century culture?

Part IV concentrates on the issue of *stability*. Monastic life was meant to be permanent: once vows were taken, there was no turning back to the world. The vows were binding and could not be legally broken.[12] Therefore, it is important to explore the virtue of stability and the sin of vagrancy, to explain the ambivalence caused by the collision of the demand of stability and the demand of being a lantern of light to the laity, since all religious people had a role in society outside their house, however cloistered they were. The last chapter is an investigation of the virtue of charity, and the role of religious' charitable activities. What were the charitable roles appropriate for religious? Were the monks, nuns, and friars represented as fulfilling these roles?

Throughout the book, it will be asked: What do people's attitudes toward monasticism tell us about their culture? How were their opinions formed? The most important question, however, is: What was early sixteenth-century culture like? When answers to these questions are found, I will be able to reconstruct a part of a world which was changing in its religious outlook, and explain why the 'Henrician Reformation', especially the dissolution of all the monasteries, was possible.

Before looking at the sources I have used, we need to take a brief look[13] at the historiography related to my work. Reformation in itself has not been one of the most popular subjects in cultural history, but the new social history certainly set up admirable analyses of religious experiences. Natalie Zemon Davis has shown us how to put the ordinary experience before the extraordinary.[14]

In England, 'history from below' unearthed new kinds of evidence about ordinary people's religion in the early sixteenth century. Robert Whiting's *Blind Devotion of the People* (1989) was very influential; he convincingly reconstructed the conservative nature of ordinary people's devotion – in his view, Protestantism spread slowly. Whiting could be called a revisionist, who set out to prove it wrong to assume that Protestantism had a profound early influence in England. Revisionists supported the idea that England remained more or less Catholic at least until the 1570s, and argued especially against the school of A. G. Dickens, who dated the Protestantization of England considerably earlier. Emphasis in the research on the English Reformation has now clearly shifted from revisionism to 'the third generation,' who stress a middle way, seek for a compromise, and admit that there were differences, both local and individual, in the acceptance of evangelicalism.

Surprisingly little research has recently been undertaken on the final days of English monasticism. The seminal work of Dom David Knowles (*The Religious Order in England*, 3 vols, 1961) is still unsurpassed. His work describes the various orders and gives a reliable account of their dissolution as well. In the field of attitudes toward the religious people, much remains to be done. In 1971, Joyce Youings mentioned that attitudes toward English monasticism have not been studied widely enough, and the situation has not significantly changed since then. In contrast, the 1980s and 1990s witnessed a stream of studies on the early sixteenth-century clergy and raised important doubts about the claims of anti-clericalism: Were people really as anti-clerical as has been assumed? This question naturally comes close to my interests. Robert Scribner, who has studied German anti-clericalism, suggests two ways of studying anti-clericalism: as a mentality and as a form of behavior (physical and verbal attacks, etc.).[15] For the present work, mentality provides more possibilities, because it is more stable than action, which can easily be affected by incidental factors such as force, persuasion, deception, lying, or propaganda. Traditionally, the existence of anti-clericalism in England has been widely accepted, and many still propose anti-clericalism as a reason for the Reformation.[16] Economic matters, especially grievances about taxation and the general failure of the Church to answer to people's needs have been offered as an explanation for anti-clericalism.[17]

The question of anti-clericalism in England seems to have divided scholars into two camps, but lately here too there have been signs of a middle-of-the-road solution to the question; that is, not everyone loved their clergy, but nor were all the clergy bitterly hated.[18] Many who do not support the idea of English anti-clericalism at all claim that anti-clericalism was nonexistent until the Protestants taught it to people. Susan Brigden's work suggests that the English were not anti-clerical before the end of the 1540s, which is when attacks on clergy began. Christopher Haigh's view supports this theory, and Patrick Collinson is also doubtful as to the culture of anti-clericalism.[19]

*

In cultural history it is important to use as varied source material as possible, and because I am tracing attitudes and opinions, I concentrate on printed English texts. Vernacular writings could reach a good number of readers lacking a formal education. I use manuscript material where it provides a perspective different from printed sources. Letters and papers regarding the dissolutions of the monasteries carried out by Henry VIII during the late 1530s shed light on the attitudes of the religious themselves toward their way of life. These papers include letters written by religious, and by the commissioners appointed to survey the religious houses. Another valuable manuscript group consists of papers relating to Star Chamber and the Exchequer. They provide information on the attitudes of people who would or could not otherwise present themselves in writing. The court cases involving laity and religious are mostly disputes over land, but they throw interesting light on various other aspects of life as well.

A coarse division of the sources used here could be the following: the first group are fictional sources: drama, romances, character literature, poetry, satire; and the second factual sources, which include moral exhortations, devotional treatises, and propagandistic prose. What is central in the first group is that the texts often repeat topics and themes well loved for centuries. As stereotypes in many of these texts, and especially satires, monks and friars are lecherous, proud, deceitful, greedy, hypocritical, gluttonous, and drunkards; nuns are depicted more or less as whorish. One of the most influential supporters of the anti-clerical theory, A. G. Dickens, presents Langland, Wycliffe, and

Chaucer as having 'eventually transmitted their attitudes to the early Tudor age, when printing swiftly magnified the impact of all three.'[20] It is quite natural that literary traditions lingered and spread, but we also need to look behind the stereotypical image of the religious person. Historical contexts can be found to provide an explanation for treating the religious as emblems of sin: for example, when their orders were established, friars caused something of a turmoil inside the Church; it was not uncommon to hear sermons against friars that had been especially commissioned by the secular clergy.[21] For many groups, friars presented a threat, and for friars, secular clergy and the members of the contemplative orders were respectively menacing, and as such a suitable target of mockery.

It must be kept in mind, however, that monks, nuns, and friars do not emerge as the only sinners in the medieval literary tradition; everybody was a potential target of scorn. Too much should not be read from their denigrated role in some literature genres. The medieval tradition reverberates, for example, in Erasmus's *Praise of Folly*, which mocks everyone. Sensibly, Herman Braet reminds us that satires and parodies were not the only tone of voice used in discussing religious people; there were times when some religious orders were reproached for having too strict a rule.[22] It must also be remembered that some of the most outrageous satires were written by religious themselves for their own amusement,[23] and for reformation from inside the Church.

My factual sources, especially many devotional treatises and studies on contemplation, were published to instruct the public. These texts were, of course, read outside monasteries as well, and used as devotional guides for all. In the sixteenth century, the numerous publications of monastic writers such as Richard Whitford and William Bonde demonstrates the need and demand for this kind of literature. Treatises composed during the medieval period were eagerly brought to print after the introduction of the printing press. The number of devotional tracts published during the reign of Henry VIII was notable.[24]

Religious literature is naturally important for my work, but cannot be clearly separated from the pamphlets and treatises. We might be inclined to categorize Thomas More's texts as religious prose, but they are quite often nothing but religio-political pamphlets. Categorizations can therefore be misleading. I have used a number of texts that remain anonymous, and hope that I will be able to give them the role they deserve. Such prolific authors as Thomas More and Richard Whitford, a Brigittine monk, tend to dominate the scene by the very volume of their work. We must let other voices speak as well, however.

As a counterpart to these, the various writings of the evangelicals form a large bulk of the material used. It is interesting that many authors engaged in debates with others whom they regarded as their religious enemies, Thomas More's texts being the best known. He answered Fish and St German among others. Against the clergy several supplications were written, from Fish to Brinklow. Most of these texts can be read as religious propaganda. Because religion and politics could not be set apart, many a political pamphlet could be read as voicing religious sentiment as well, some in the more conservative camp, some in the evangelical. From there it is not a long way to the propagandistic texts produced by Thomas Cromwell's men. Of the pamphleteers that were influential after 1536 – the watershed being the first dissolutions of the monasteries – there are several that should be mentioned. Two of the official apologists of Thomas Cromwell, who was appointed by Henry VIII vicegerent in religious matters, the humanists Richard Morison and Thomas Starkey, wrote especially about obedience toward the king.

Regardless of what has been said above about various known authors, it is in fact less important *who* wrote what, than *what* was said and *how*. It is painfully obvious to me that even though I have tried to read all kinds and types of texts, and as many as possible, I have ended up with results that are wanting. The authors of the works cited are mostly highly learned men, sharing the same experiences: they often went to the same universities, traveled in the same countries, met the same people; and they were clergymen and lawyers. They were members of the learned society for whom it was natural to express themselves in writing. Consequently, the material at hand seems to tell a story of London-based, 'upper-middle-class' males. I trust, however, that drawing general conclusions from my material is possible on the basis of their collective nature – most of the attitudes presented can be found in several sources. History without the possibility of generalization would be pointless. Furthermore, variation from this dominant group is offered by poets, the letters of the religious, and the court cases mentioned above. This means that the images that my sources form are not necessarily one-sided.

A short note regarding terminology is in order here. The term 'religious' always refers to someone who has professed, that is, someone, whether contemplative or mendicant, who has made the monastic vows. The term 'monastic' refers to members of the contemplative orders. Furthermore, I have decided to reject the anachronistic epithet 'Protestant,' since at this point, people were not yet Protestants. They might be Lutherans, Anabaptists, or Lollards, etc., but not yet Protestants. People had various opinions, some of them very much influenced by the new learning, and many called themselves or others *evangelicals*. I will make an effort not to categorize people of the past if they do not do that themselves. Those who supported the old religious order I call Catholic or conservative.[25]

*

It is self-evident that the early sixteenth century, a period of religious change, needs a new kind of scholarly approach. This study sets out to approach the cultural history of the age of Henry VIII from a fresh viewpoint, stressing the importance of understanding the culture of that age in a wider context. I am convinced that when we comprehend the ways in which religious attitudes were constructed, we are better able to identify what made historical changes possible, or indeed perhaps suggest that continuity was actually more important than change.

Probably the greatest change – at least in religious terms – that took place during the reign of Henry VIII was the dissolution of the monasteries. Scholarly tradition has primarily explained this on economic grounds: religious houses were suppressed simply because the king needed money. This is not difficult to accept, but the question of how the dissolutions were possible if people needed religious houses has not been convincingly answered. I believe that the answer lies in the way people thought.

Part I
Obedientia

1
Obedience – Renouncing the Will and Pride

> Adam of erthe sone of vyrgynyte
> And Eue by god of Adam create
> These .ii. the worlde dampned in certaynte
> By dysobedyence so foule and vycyate
> And all other than from them generate
> > Hawes: *The pastyme of pleasure*, sig. T1

The virtue of obedience formed a scaffolding around sixteenth-century society; it maintained hierarchy and supported the structures essential for stability. Obedience provided all Christians with a sense of belonging. Since obedience maintained order, no society or ruler neglected efforts to ensure that obedience was a virtue respected at all levels of society. From homes to courts, obedience maintained hierarchy.[1] All disobedience, whether open resistance or more passive reluctance to do one's duty,[2] was in the end regarded as directed against God and Christ. To obey with humility was the duty of a good Christian. Only dissolute men ensnared by sensuality ignored Christ's call to obedience – and were doomed.[3] It was stressed that obedient people were Christ's beloved children. He denounced all others, who then became sons and daughters of Lucifer. The disobedient were like rude pagans for whom the great child of the devil, the favorite example of disobedience of the time, King Nebuchadnezzar, had set an example of a life so horrible that it defied reason.[4]

Obedience served as an agent of eternal life, disobedience as that of death and destruction – it was the very first sin of humankind, and the curse of disobedience was a fruit of Adam and Eve's sin.[5] What could teach obedience as efficiently as constant reminding of the eternal bliss humankind had lost because of those miserable two? It was painfully clear that obedience was the only known antidote to original sin. Only obedience to Christ and secular power could bring redemption, and eventually, with the help of Christ, restoration to a sinless state.[6] With this in mind, it is not difficult to understand why the vow of obedience was often considered the most important in religious life.

Medieval obedience was very much a monastic virtue.[7] In religious life the essence of the vow of obedience was complete, voluntary abnegation and renunciation of one's own will. When a monk or a nun entered the religious life, he or she voluntarily surrendered his or her own will, because he or she was no longer supposed to have any need for it. Similarly, religious people gave all their worldly possessions to the monastery. Power and decision-making in religious houses rested with the head of the house. The monk's will was given to his superior, the abbot, who would exercise this will. In an ideal case the abbot acted according to Christ's rules and led the collective righteously. In this way the abbot or abbess would not actually use his or her own will at all, and could thus remain faithful to his or her own vows. The greatly revered St Bernard of Clairvaux, among others, had written that a religious person had to make sure that it was the will of the head of the house which was followed, not his or her own.[8] Giving up one's will was difficult, however, and great striving was needed. A synthesis of the late medieval ideal of the religious life can be found in the extremely popular *Imitatio Christi*, a best-seller of the later Middle Ages. Monasticism meant breaking one's will, throwing oneself outside the world so profoundly that one could experience delight in being regarded a fool. This was mortification, the true medicine for the soul. The monk is a monk because he feels the need to serve:

> What shulde be the lyfe of a true relygious person. It behoueth the to breke thyne owne wyl in many thyngs if thou wylte haue peace & concorde with other.
>
> If thou wylte stande surely in grace & moche profite in vertue / holde thy self as an outlawe and as a pylgryme here in this lyfe / and be glad for the loue of god to be holde as a foole and as a vyle person in the worlde as thou arte. The habite and tonsure helpe lytell / but the chaungynge of lyfe and the mortifienge of passyons / make a person perfite and true religious . . . It is good therfore that thou remembre oft that thou camest to religion to serue and nat to be serued / and that thou arte called thider to suffre and to labour / and nat to be ydell ne to tell vayne tales. In religion a man shall be proued as golde in a furnace / and no man may stande longe there in grace and vertue / but he wyll with all his herte meke hym selfe for the loue of god.[9]

It is important to notice that the perfect obedient state could be reached only through action: the mind had to be trained as well as the body. Mortification was needed for both. In later chapters we will return to the practices of mortification in more detail.

The abbot and the abbess had a somewhat different relation to obedience than ordinary monks and nuns.[10] They, too, had vowed obedience, but had been later elected into a position in which they had to exercise the will of perhaps dozens of religious. The duty of the head of the house was to see that his or her authority was never questioned. All these religious shepherds were advised to be careful and watchful for rebels in their flock. The well-being of the flock (well-being was often simply a synonym for discipline) was entirely in the hands of the monastic heads: disobedient monks and nuns endangered their brothers' and sisters' chance of a divine life.[11] At worst, disobedience could spread like a forest fire and bring about the destruction of all souls. The monastic rule was designed to prevent this from happening, to keep the religious spiritually and mentally contented and submissive. For religious, the only will that remained was the will to serve God and obey his commands. This in its turn meant obeying his commands and the monastic rule.[12] The word of the head of the religious house was not only to be obeyed; it was fundamental for a monk or a nun to *want* to obey. True monastics were exhorted to commit themselves completely: '[i]t behoueth the to breke thyne owne wyl in many thyngs if thou wylte haue peace & concorde with other.'[13] The essence of monastic life was renunciation: giving up one's past, family, possessions, and one's will. Of these, giving up one's will was the most meritorious, as it was perceived to be much more difficult compared to relinquishing one's mere belongings: 'sythe the forsakyng of oure owne wyll is moche more noble / than the forsakyng of worldely goodes,'[14] wrote William Bonde, himself a monk.

When successful, obedience worked in miraculous ways. It was the health of the soul, nourishment of virtues, and the gatekeeper of heaven. It brought heavenly pleasures for the believer, and was accompanied by sweet perfection:

> Obedience / saithe one other of the holy fathers / is the helthe of faithfull soules / the nourse of all vertue / the fynder of the right waye to heuyn: It opyneth the gate of heuen: it lyfteth man form the depe pytte of synne / and daunger of this worlde. Obedience is the neighbour of angels / the fode or meate of holy sayntes: for by obydience they were wayned frome this worlde / and fro the pleasures of the same / and therby they came to perfection.[15]

For the monastic who pursued the angelic life, *vita angelica*, it was comforting to know that obedience was perceived as the neighbor of the angels, and would give him or her a good chance of becoming an angel. It was a noble and logical goal for the monastic life.

Sixteenth-century theology distinguished between different forms of obedience, which all offered the religious person some merit, but differed as regards the amount of spiritual good they would yield. According to one classification, the first form of obedience was motivated by pleasure or profit, the second by need, and the third by fear of pain or punishment. The last mentioned certainly was the most efficient. Religious were reminded that obeying for pleasure was not as meritorious as obeying when it was difficult, unpleasant, or even repugnant. As always, hardship generated weightier rewards.[16] Ideally, monastic obedience should be liberal and loving, strong, mighty, and bold, and it was always to be realized with due reverence.[17]

In a sense, extreme obedience can be interpreted as marginalizing the monk. Building on Ephraim Mizruchi, Barbara Hanawalt notes that in the margins there can be established groups such as monks or apprentices. Importantly, she points out that margins could be defined by the space the people in the margin occupied. Thus monks formed a marginal group just as prostitutes did.[18] Like Hanawalt, I believe that spatial marginalization made the boundaries between the center and the margin even stronger. Inside a cloister, by giving up his will, the monk surrendered a significant part of his humanity and personality, which made him unfit for the world or ordinary life. It is highly probable that he was even further marginalized, because submission and complete obedience were socially and mentally, as it were, feminine concepts. 'Feminine' was thought to be by the law of nature less than 'masculine.' For example, an allegory of obedience could be a good servant who obeyed her mistress, despised all vanities, and was fully content in fulfilling all her mistress's commands. Disobedience, on the other hand, was a wanton woman who resisted and defied her mistress. She was vain, and might even tell her mistress that the mistress was wrong.[19]

In lay society, obedience was linked to the fourth commandment – children had to honor their parents.[20] As we have seen, obedience maintained the hierarchy. Furthermore, obedience was often seen as a womanly sphere. Anything feminine was always lower than masculine and it thus followed that women and the feminine were lower than men and the masculine.[21] This notion was enforced by women as well as by men. For example, Christine de Pizan had stressed the good womanly qualities of obedience in marriage as well as a nun's obedience in her convent.[22] Good wives were obedient. Natalie Zemon Davis examines the gender-specific nature of this virtue, and shows how a division of virtues worked in court cases dealing with the murder of husbands and wives. Men's motivation would be the disobedience of their unfaithful wives, but not, for example, jealousy.[23] Proof of the poor quality of female obedience had already been received in Paradise, where Eve had chosen not to obey. Now, in order to regain the original sinless state, all women were inferior to men – or at least to their fathers and husbands. There were, of course, exceptions to this rule: women were heads of their servants, and their children were to obey them. Luther was one of the strongest proponents of this ideology.[24]

It is not suggested here that obedience was not an important virtue for everyone; it is merely proposed that a monk's obedience marginalized him by setting him below other men. It made him peculiar in a marriage[25] to Christ – and the lower was expected to be the submissive one. Submission minimized his role in all aspects of his life, for it took away his sense of control. This was, of course, the monk's intention, but from the vantage point of laymen, monks must have seemed a peculiar group of men. Although all laymen had someone to obey, they never renounced their will. The monk's obedience set him apart from the outside world; in this sense, it was a double seclusion. No doubt this peculiarity as regards obedience was intended to add to the holiness of the monks and nuns, and raise them above ordinary humans.

The importance of obedience was repeated over and over again, and religious seem to have understood the meaning of this virtue as a central element for the success of religious life. The poet John Lydgate offers us an account of his life, which was a long journey from evil toward religious perfection. It exhorts people not to cherish outward matters at the expense of the soul's well-being. His *Testament* is in the English context a rare example of a confessionary account, a recollection of a monk's process of becoming aware of the need of religious repentance through contemplation.[26] Lydgate wrote about his utter disobedience and the manner in which he came to his right mind and found the light of Christ. Like St Augustine, Lydgate had been a man of the world, but, almost miraculously, had found the true meaning of life and religion. Lydgate confessed that as a young man, before entering a monastery, he had led an evil life. Not even the monastery changed him completely: he worked hard in the fields, and was duly professed after a year as a novice, but he still yearned to be back in the world. He compared himself to Lot's wife who proved her disobedience by looking back at

Sodom and met a horrible fate.[27] In this sense Lydgate makes clear that
regretting the loss of the world was utter disobedience.

Even after his profession Lydgate was disobedient; he recognized the
right path but did not follow it. A sinful life was easier, and changing
direction took time. Lydgate even wrote how he would rather steal
apples and grapes from gardens than attend religious services. He confessed that he had good, virtuous, and religious teachers, but took 'but
lytell hede'[28] of their sound advice: 'What I was bydden / I coude well
disobey.'[29]

Lydgate's change of heart was dramatic in its simplicity. A crucifix on
a cloister wall changed him, and seeing the wounds of Christ made him
realize his sinful state. By the crucifix were written the words: 'Beholde
my mekenes chylde / and leaue thy pride.'[30] That moment made Lydgate realize the full potential of the life in Christ. It is not accidental that
Christ asked him to leave his pride since one of the greatest threats to
obedience was the sin of pride, one of the two arch-sins (the other was
idleness), because it unavoidably led into disobedience. Because of
pride, thousands of souls suffered the torments of hell.

The sin of pride was regarded as truly horrible, since it was the very
reason for the fall of the angels, of whom Lucifer was the most
frightening case. Similarly, it was thought that it was pride that led
Adam and Eve into their fatal disobedience.[31] In his treatise on passions
Vives pointed out that pride was 'a monster composed of many emotions, joy, desire, and confidence,' and noted its insidiousness: 'Pride
creeps into our soul through little underground tunnels. It can often be
caused by our great virtue.'[32]

Pride was indeed deceptive, as it could lurk anywhere. In the war for
souls, humility was an essential weapon, and for a monk especially,
because pride in those who had nothing was far more despicable than
in great men.[33] For many early modern people the worst monks were
not those who were mad enough to think they were in heaven, but
those who judged others. Again, traces of the marginalization of religious people can be seen here, since their own ideals gave them very
little room to operate.

Even though pride was a very mundane sin, not all pride was caused
by worldly things; spiritual pride existed also. It could spring from
excellence in religious life and countenance, and it was regarded as
extremely dangerous to the soul. It developed simply from thinking
too highly of oneself. First, the spiritual person would begin to neglect
God; second, to judge God; and third, this person would reach a state
where he or she would begin to disobey God and rebel against him.

Ultimately, this meant forsaking God altogether.[34] Losing the battle was easy; anyone who excelled spiritually could easily lose God's favor if he or she did not keep a close watch on his or her soul, and stay humble when the top of the ladder of religious excellence was reached.

Obedience not only conquered pride; Whitford noted that it also defeated envy and wrath, and promoted meekness, charity, kindness, and patience.[35] Consequently, the paired concepts – obedience and pride – were linked to all other sins and virtues, and similarly, they related to other monastic vows. It is therefore no wonder that the sin of pride is so prominent in texts exhorting the reader to religious excellence that it seems to dominate these texts. Pride was found lurking in diverse actions and feelings of an unexpecting monk: *Shepherd's Kalendar* listed vainglory, gladness of doing evil, boasting of sin, disdain, tempting God, excess, dispraising, false goodness, hardness, presumption, rebellion, obstinacy, and sinning wittingly.[36] Pride was so very difficult to avoid because it was so treacherous. It could reside in even the minutest detail. This is why it required relentless vigilance. Many devotional texts regretted that pride could sometimes be the reason for entering religious life. It could emerge in the search for personal praise and merit, or in the acknowledgement of one's great religiosity. Pride and ambition in a monastery were a lethal combination; not only was the ambitious person himself affected, he was poison that tainted 'moche good meate.' Eventually, he would destroy the house.[37] Consequently, in a religious house of truly obedient sisters or brothers there was no place for personal achievement. All forms of competition in the amount of devotion and religious fervor were to be abhorred. The mind had to be free from sentiments of ambition; humility had to take their place.

Humility, the efficient counterpart to pride, was desperately needed to repay the debt to God, because everything a man had, including his lusty body and his sinful mind, he had received from God.[38] As repayment, God deserved better than sin, for his love was too far-reaching to be thrown away. Humility was efficient when expressed by a genuinely penitent devout. God was just: his forgiveness was granted even after very simple exercises of humility. For example, if the monk made a mistake when he sang or read aloud, he would apologize and express his humility in a manifest way, stooping down to the floor.[39] Common practices like this encouraged the development of virtues and gave protection from devilish thoughts filled with pride.

People understood that even the most devout would occasionally fall. This happened to a very virtuous nun in a story told by Thomas More. A doctor of divinity visited his sister, a nun, after a long absence. The nun was known for her virtue, and her excellence had made her brother very proud of her. The order in which the sister was professed was very strict, and the two could meet only at a gate and touch only with their fingertips through the grille. As soon as they met, the sister began to talk, and she spoke incessantly about the dangers of the world without recognizing the extreme rarity of the occasion or treasuring the chance to be with her brother. Finally she paused for breath and asked her brother why he did not give her any spiritual guidance, since he was, after all, highly learned in divinity. The brother, apparently exhausted, could do nothing but sigh that he had not been given a chance to open his mouth.[40] To a modern reader the brother seems quite as arrogant as the sister, but this seems not to have been More's point. The one who got his sympathy was the brother. The nun was left without spiritual guidance because of her pride in spiritual matters: she did not appreciate the presence of her brother enough, and her conduct revealed that she was not prepared with suitable and virtuous humility for the rare occasion of her brother's visit. Furthermore, another weakness can be detected. The sister was, regardless of her virtue, unable to rise above her womanly faults. It was a common perception that women babbled incessantly. As Natalie Davis has shown, women who expressed strong opinions were sometimes depicted as monstrous.[41]

Pride and humility were issues that interested all parties in the religious debate during Henry's reign. It is therefore no wonder that among other sinners, religious who sought to enhance their own honor and to rule others caught the eye of the writers and captured their pens: '[they] procure theyr owne honour, and call it the honour of god, and rather couet to haue rule ouer the people, than to profyte the people.'[42] Religious people were not only accused of being proud, their relationship to God was also criticized as being twisted. The services of the proud religious were seen as distancing them from God rather than bringing them into union with him. This argument, as we can see, is not far from the reasoning of the religious themselves; the critics thought that pride was the prevailing sin in religious houses.

Quite cleverly, Thomas More turned the accusations against religious against their accusers, who, he thought, were themselves gripped by pride. In fact, he saw the whole religious turmoil as caused by the pride of the evangelicals. He found support from St Augustine, whom he cited: 'pride is . . . the very mother of all heresyes.' All in all, pride was so complex and yet so general in nature that it was not difficult to proclaim the heretics sinners in pride, or as More did, apes of Satan, who sought only people's acceptance and worship. The heretics pursued the perfect life in vain, and gained nothing by their illusions. They were the devil's martyrs, who lost everything trying to please their horrible master.[43] More was, however, fully aware that religious people were sometimes plagued by pride, and he detested that. As we saw in the tale of the proud sister and her brother, More wrote *exempla* about spiritual pride and utterly denounced it. One of these, almost a horror tale borrowed from Cassian, was about an anchorite whom the devil tempted into committing suicide after convincing him that an untimely death would take him straight to heaven.[44] More unforgivingly reproached the anchorite and underlined that the man had lost everything, his life and, what was worse, his soul, simply because of his foolish pride.

The sixteenth-century mind saw spiritual pride as blinding reason and good judgment. Reformers denied that pride would best be conquered in religious houses, because the monk's or the nun's faith was plain, arrogant pomp, a useless recital of words, 'great gapynge and openynge of the mouthe,' vain teaching, and profited no one.[45] To assert that people in general thought that traditional faith was mere gaping of the mouth would certainly be naïve. Similarly, there is no reason to expect people not to have been aware of the pursuit of reform inside the Church itself. People knew that there were problems in religious life; nor did the religious themselves pretend that everything was perfect. Quite the contrary, there were critical monastic voices, for example, condemning attempts to enter the monastic life for the wrong reasons. Among these voices was Richard Whitford, who attested that a poor background meant being especially susceptible to pride when suddenly surrounded by monastic wealth. Such a person would easily act as if he or she was of noble blood: complain about the food, become fastidious and a great connoisseur of food and wine. Such a monk or a nun would flatter the abbot, and react badly if his or her behavior was in any way reproached.[46] Whitford's words reveal not only his disapproval of pride in religious houses, but also the contemporary attitude toward hierarchy. It would be unwise to assume that a monastery was a democratic community which would challenge the rules of the hierarchical society outside its perimeter walls. That was not the case. Those who aspired to cross the boundaries (formed mostly by birth) were scorned, like the beggar who arrived at the monastery gate 'hauynge neyther a good shurte nor coate' but who, suddenly when given decent clothes, became a gentleman, 'though he be but a starke knaue.'[47] Without doubt, pride was a question of class and origin as well. In some cases, and especially in the case of the poorest in society, choosing the religious life as a vocation could be read as an act of pride in itself, of not being humble and satisfied with one's origins. This was not always the case, but the seed of suspicion seems to have been there.

The overall constructions of society were visible in the structuring of religious communities. They were a part of the society and their faults and virtues reflected the world outside. Looking at the ways in which sixteenth-century English people perceived their superiors and God, it is not difficult to understand what a major role obedience played in the way they constructed their world. The very structure of the society that surrounded them daily tacitly maintained the hierarchy and the need for obedience. There was always someone above you, and obedience and submission were needed for survival. Those who were high up on the ladder had to be honored, but even they had to submit – to God – if they wanted to save their soul. Similarly, the relationship to God was recognized as being largely built upon obedience, humility, and recognition of one's place in a hierarchical society. God was best served humbly and contentedly, and by not asking for something undeserved.

In the 'margins' – even though in the case of the monastics, the 'margin' was one filled with potential for power – of this hierarchical society, monasteries maintained the image of true obedience. This virtue was always a central part of the popular images of religious: for their critics it appeared in the form of the sin of disobedience, and for their admirers the ideal monk or nun was the personification of obedience. Admittedly, obedience was easily broken, but if maintained, it could produce a coveted place among the angels. Obedience put religious in a position in which they were, at least theoretically, the lowest and humblest of the earth. They were thus marginalized and separated from ordinary people. In these margins they were occasionally left to themselves to strive for what had been generally understood to be a noble cause. Ideally, obedience was the greatest singular achievement of the true monastic, the very virtue with which he or she could, so to speak, stand upon the wall and shout to the world a message of monastic excellence and superiority. And this, of course, was dangerous. Emerging from the margins triumphantly was a hazardous business.

The ideal monk or nun was presented as a creature whose own mind was emptied of worldly matters. The only will he or she had was toward God. In many ways the perfect religious was an image of a perfect citizen, since his or her own will was only to obey. In contrast, the image of a rebellious monk or nun symbolized disobedience in the wider society. This is what we will discuss next in more detail.

2
Questioning the Power of the Sovereign

As we have seen, obedience was a key to religious life and for everyone a key to eternal life as well. It was a supportive structure of society, the stonework in the walls of a monastery that kept vices out and virtues in. Obedience was also the pillar upon which secular rulers built their power; if it crumbled, there would be little left to rule. Next, the focus will be momentarily shifted from inside a religious house to obedience in political thinking and everyday politics. This will be done in order to illuminate the structures of power dominating the political and religious scene during the reign of Henry VIII in relation to the monastic question.

It is perhaps not unreasonable to ask why the king's power was relevant to the issue at hand. Everyday politics during the late 1520s and the 1530s especially do not at first glance appear to have anything to do with attitudes toward monasteries. First, however, understanding the nature of a significant number of sources written at that time requires an understanding of the ways in which political theories were developing following the process during which Henry was declared head of the Church of England. Second, the fate of the religious houses was dependent upon Henry's politics, since politics and religion could not be separated in the sixteenth century. Understanding religious ideas requires taking political ideas and events into consideration. Third, and most importantly, obedience was a virtue which permeated society at all levels. Religious obedience was in a direct relationship with worldly power in two ways: on the one hand, the religious idea of obedience influenced political thinking, and on the other, religious obedience was obliged to take lay power and its requirements into consideration. The following is an analysis of these relationships.

The beginning of the formation of the medieval and early modern English monarchy can be traced to the Norman conquest of 1066.

Although William the Conqueror was prudent and not willing to force strong foreign feudal practices on the English regime, it was his influence that brought all English land – at least symbolically – into the hands of the king.[1] This is a significant matter (even though the king's power over all land was not always recognized as immediate), because when a king wanted to confiscate land, it was, as it were, his to take. This is a right that Henry and, on a minor scale, many of his predecessors used when they wanted to establish new institutions, such as schools or colleges. Rosenwein posits that since the reign of Edward IV monarchs had 'increasingly claimed to control the nature of English space.'[2] Consequently, the king's rights were extended, and monastic lands and property had often been taken into royal use if there was need for new institutions or foundations. Dissolving religious houses was not Henry's or Thomas Cromwell's novel idea.[3]

The king's right to the land was not just a historical feudal right without deeper justification. On the contrary, his sovereignty was based on the notion of its divinity: the source of the king's extensive power was God. What happened in the ideology of the monarch's power during the centuries from the early medieval electorate kingship to the peculiar kingship of the Tudors, which resembled a quasi-cult, was a long process. It has been explored by many distinguished scholars[4] and we will take only a brief look at the issue here. At the end of this process, the king had become in metaphorical terms untouchable, and all obedience was to be directed to him. This meant that even a disastrous king could not be deposed.[5] Theological confirmation of this untouchability could be found in the Bible, in which I Samuel 26:9 stressed the untouchable nature of kingship. Of equal importance were St Paul's words to the Romans (13:1) in which he commanded everyone to obey the ruler that God had given to them. This was a key issue, and it was in constant use whenever it was time to prove that unquestioned obedience belonged to the king, and not to the pope.

Whenever further evidence for strengthening the king's power and its divine character was needed, it was readily available. St Augustine had written that God alone could give realms for men to rule, but he could give them to either good or incompetent rulers according to his unerring wisdom.[6] Therefore, the people had no choice but to obey the ruler they were given, evil or not. An incompetent king could thus be seen as a deserved punishment for sins, and gave no room for rebellion.[7] Uprisings were thus regarded as directed against God. They originated in Antichrist, and all rebels were his children, no matter what the cause for their discontent. Nevertheless, some important medieval theologians stressed the contrary and argued, like St Thomas Aquinas, that people had a right to rise up against an ungodly prince. St Thomas added, however (certainly to the relief of many rulers), that because rebellions were not profitable to anyone's soul, unlucky subjects should turn to prayers instead of earthly weapons.[8] In the sixteenth century, these aspirations could be turned in and focused on a direction favorable to the king and his regime.

Kingship was sacred and eternal. Because the king's power was divine, anyone who offended him was doomed. This was the core of the untouchable nature of kingship.[9] Immortal kingship is discussed in detail by Ernst Kantorowicz, whose work explains the communal nature and the divinity of Henry's crown. The communal nature of kingship indicated dynasticism: a hereditary monarchy eradicated the older form of kinship, which had been based on the agreement of the ruler's subjects. When Henry VIII's son ascended the throne as Edward VI, he had divine power behind his rule even without the blessing of the Church.[10] Power had been given to the Tudors and it was to remain theirs forever. Jennifer Loach argues that Henry's coronation emphasized the king's sacredness – the monarch was a saint.[11] It should be noted here that what Marc Bloch demonstrated in *The Royal Touch* about the miraculous nature of kingship still applied: people continued to believe that the king's hand could perform miracles and heal the sick. This suited Henry well, and it was his politics to continue the tradition. He ordered Thomas Cromwell, for example, to distribute traditional healing cramp rings in his name. As an anointed king, Henry was certainly thought to possess the healing touch of God's chosen. These rings symbolized his good will and, more importantly, elevated his magical and divine qualities to the spheres needed for his credibility.[12]

The pope's power was unavoidably undermined as the king's power increased, but this was no novelty. The role of the papacy had been under constant change and pressure. David d'Avray views the period of glory of the Roman pontiffs as very short, culminating in the later half of the fifteenth century.[13] Whether greatly honored or not, the pope's power was always treated with caution. The papal bull *Unam sanctam* issued by Boniface VIII in 1302 had defined the pope as the ultimate spiritual authority, who could not be judged by anybody, but who judged everybody.[14] Boniface's initiative was not, however, accepted by all.[15] Disagreement on this question is not surprising, since there was no such a thing as a monolithic church observing strict orthodoxy. As R. N. Swanson, pointed out the devotion of the medieval Church was 'demand-led by the spirituality and desires of the laity. That, of course, provoked a good deal of uncertainty within the institution, and weakened its ability to present a unified, monolithic orthodoxy.' Swanson notes, moreover, that religion was neither purely institutional nor purely theological: it was practical, with an emphasis on '*living* the Christian life.'[16] The practicalities of everyday devotion often outweighed questions such as those discussed above.

It must be remembered that although the divinity of Henry's rule was promoted continuously from his accession,[17] he was not a priest, and it was not until the Act of Proclamations in 1539 that he received the right to amend the teachings of the Church of England. After some debate, Henry was not granted priestly rights to preach, administer sacraments, or consecrate bishops.[18] Many lawyers thought that although it would be legally possible to have a priest-king, being a priest would in fact not increase the king's honor in any substantive way.[19] It was better that he remained a layman, and this view prevailed.

Henry truly aspired to be a philosopher-king. In this he followed humanist ideals. Not only was he well read in theology, he was famous for his humanist upbringing. He studied the classics and divine works to seek spiritual greatness. He was taught that he had to be just, live in virtue, be good, benign, God-fearing, and honor his subjects.[20] Such a king could aspire to God's approval, as not even a king was without a superior. Salvation is guaranteed to no one; the disobedient will certainly perish:

> He that mekely will not obaye
> Shall perish wyckydlye it is no naye.[21]

In practical terms, Henry's kingship was politically consciously constructed. Scholars often propose that Henry's reign was marked by a strong fear of uprisings and conspiracies. Such fear was a creation of the restlessness of the early years of his reign. Throughout Henry's reign there were uprisings and prophecies undermining his claim to the throne.[22] This is relevant to the study at hand since Henry's wariness had an effect on religious houses. It explains, for example, the deep feeling he had about the cult of Thomas Becket, who was now considered a great traitor.[23] He feared sedition, and when the monasteries were being suppressed, Henry had already acquired considerable experience of handling dangerous situations. If they had not already done so, his subjects were to learn that the king was a ruler with powers given from above. If he was fearful, his fear did not manifest in irrational actions.

Perhaps because power is always dialogic,[24] Henry launched a campaign on several fronts propagating his supremacy to ensure success in the battle with the pope. It commenced early, in the course of the process of the divorce from Katherine of Aragon. Steven Haas dates the beginning of the campaign to 1531, on the grounds that 1531 saw the first pamphlets on the issue distributed in Parliament and the convocation. The early campaign was constrained, since bridges with Rome were not meant to be completely burned. In domestic terms it was recognized that it was important not to alienate the people.[25] The king's supporters stressed that Henry was sovereign by the will of God, and that obedience to him had to be absolute since divine law allowed no disobedience. Like religious authors, pamphletists noted again and again that obedience was the mother of all virtues. It was Christians' God-given duty to be obedient to the king. [26] Rex quite convincingly argues that the question of obedience was the very essence of the Henrician Reformation, but he dates the propaganda campaign on obedience later than Haas. The campaign was especially forceful during the later 1530s.[27] After the final breach with Rome, it was stressed in the propaganda campaign, which now produced dozens of treatises, that if obedience failed in England, disaster would follow. It was genuinely feared that England would be invaded by the pope's troops, and that this would be the end of the nation. Cuthbert Tunstall wrote in his *Sermon* that England would become 'a praye to al venturers, al spoylers, al snaphanses, all forlornehopes, all cormerantes, all rauenours of the worlde.' After the invasion, no one would be safe: wives and daughters would be raped before their husbands and fathers, children killed, houses destroyed, cattle, plate, and money accumulated for children plundered. After which, anyone left alive would be slain. The only person who could save the country was Henry, who was 'hardy as a lyon.'[28] By paying homage to him and obeying him, the realm and the lives of thousands would be saved.

Understanding this atmosphere is important for the purposes of the present study. The careful control of publishing, the systematic spying for any papist words, censorship, and oppression by executing the supporters of the pope, generated fear and silence. It became very difficult and extremely dangerous to speak in support of anything linked with the popish religion, including religious houses. Even private conversation could count as treason if one was caught.[29] Times were hard and survival in both groups – in the camps of Catholics and reformers alike – could be equally haphazard. What is called the Henrician Reformation has at times seemed quite ambiguous, but recent work on the development of

Henrician theology has shown that there was more consistency to the king's policies than has been previously supposed.[30]

The 1530s and especially his famous first divorce can be seen as the culmination in the development of Henry's power. The debate on the location of divine power was especially heated, and the question of supremacy was inescapably intertwined with the process of the divorce.[31] In the end, it became clear that the only way to obtain a royal divorce was a grand divorce from the Roman pontiff. Since obedience was such a fundamental issue, it was imperative to redirect obedience and loyalty, which necessarily followed obedience, from the pope to the king. In this undertaking, the medieval idea of kingship, power, and the legitimacy of rebellion worked to the king's advantage. Somewhere in the middle, or just beneath this discussion, was the fate of religious, who on the one hand became unavoidably involved in the politics of the day, and, on the other, were largely ignored when they expressed their opinions regarding the dissolutions.[32]

The downfall of the monasteries, when it came, was rapid. The Act of Dispensations in 1534 had severed the links between the English monasteries and the pope. All religious houses were now 'national,' and monks were no longer allowed to appeal to their continental sister or mother houses.[33] Felicity Heal has noted that the first rumors about the king's intentions to confiscate church property and to redistribute it to the nobility were circulating in January 1534.[34] This murmuring was probably founded on the visitations for *Valor Ecclesiasticus*.[35] The relationship is not without logic: the *Valor* was completed in 1535, and the dissolutions of the smaller houses began in 1536. Because many houses which were shortlisted for dissolution were allowed to continue, it seemed that the dissolutions would end at that and that the process was not much more far-reaching than earlier dissolutions had been. But in 1539, when the Act for dissolving the larger houses was passed, it became evident that the king was in earnest, and that all houses were to go. There is no denying the fact that his motivation seems to have been largely financial.[36] But how was such a complete suppression possible, and why? Fundamentally, it *could* happen because it was a question of ownership of the religious houses. Traditionally, the monarch had been the founder of many houses, and the king's feudal rights over all land provided a legal basis for his actions. He could do what he wanted with something that was his own. Moreover, if the dissolutions were aimed at reforming religion for the well-being of the nation, the supreme head's will overruled other wills. Politically, the maneuver was skillful. Once Parliament had passed the Act of 1539, there was very little that could be done to save the monasteries.[37] The question that needs answering is why the English allowed religious houses to be destroyed. To answer this question, we need to investigate what was thought of the dissolutions and how people thought and formed their attitudes. Simply put, we must ask if their attitudes had a role to play in these events.

3
Monastic Obedience and the Religious Houses

In the tumultuous political and religious climate of the 1530s, religious houses came to be perceived as a threat to the king's policy of reform. As a practical consequence of the Act for the Submission of the Clergy, all clergy were bound to obey only the king, not the pope. Traditionally, many religious orders had had close ties to continental mother houses and more or less directly to the pope as well. The loyalty of these so-called 'alien' houses now became suspect. Naturally enough, many felt that pro-papal sentiment would linger in religious houses and perhaps increase unhindered. This fear is manifested in many texts which represent monasticism as a bottomless pit of violence, viciousness, and conspiracy. For political reasons it was useful to maintain that monastics held awesome power, that they were horrendous monsters who kept princes and emperors and, indeed, even the pope in great fear.[1] Because fear created power, in the most fantastic of ideas presented it was the religious who ruled the world.

These accusations, even the most extreme, are worth examining since they reveal interesting thought patterns and strategies in the religious struggle. Below, we will examine the ways in which power met and evaded the religious. We will also look at the ways in which attitudes were reinforced, transformed, and even fundamentally changed. The main topic here is power and religious during the 1530s.

Leaving the accusations of Henry's supporters aside, it must be noted that it was usually accepted that religious were more commonly subjects of power than executors of it. To fulfill the most basic of their obligations they had to be obedient to the pope. Success in revering the Church had traditionally brought glory to priests and religious,[2] but when the loyalties of the government shifted radically against the pope, monastic obedience toward the old regime and clinging to the old ways in religious houses became suspect and an obvious target of scorn and denigration. But, it must be added, monasteries – which were on the margin and yet powerful – were also sources of fear.

Denying the pope's unique position required redefining him as the fount of many an evil. Consequently, he was depicted as a servant of Lucifer, and other maintainers of the old religious order, such as cardinals, bishops, monks, and friars, were viewed as branches of the same tree.[3] The whole old order was of Antichrist. The pope's rule was seen as a long line of horrible errors in which the clergy with their misguided followers lingered on in evil. People were kept blinkered by constant preaching. Obedience was seen as beneficial, but when it was directed to the pope's rule 'nyght and daye,' a good thing became evil and a source of inevitable destruction, because in reality reverence was paid to the Devil.[4] The pope's title was God's vicar, but for some he was nothing but the servant of Antichrist.

An individual monk or nun could lead a religious life without constant consideration of the loyalties and structures of power outside his or her cell, but as a group, monastics were daily affected by power and the question of worldly loyalties. These expectations and obligations emerged from different directions, and they were varied in their outlook and meaning. The outside world penetrated the monastery walls at many levels. First, religious houses were almost always integrated into the communities that surrounded them, especially in rural areas. In Bury St Edmunds, an entire town developed around a monastery. The relationship of a monastery and town could be a dialogue consisting of trade, employment, religious services, and so on. On the one hand, Bury St Edmunds was dependent on the existence of the Abbey, and on the other hand, the Abbey could not have survived without the town. As Robert Gottfried has noted, the abbey was 'essential to the very fiber and being of the town.' The dialogue continued until the Black Death, when the town gradually became independent and self-supporting. These changes meant that the monastery was largely left to its own devices, even though it remained an important consumer of the goods produced in the town.[5]

Second, the fate of religious houses was directly dependent on the local aristocracy. Because of their rivalry, a strong aristocracy often meant a weaker monastic influence. If the influence of laity was exceptionally strong, as it was for example in Cheshire, monasticism could not flourish. This was so for at least two reasons. A strong earl prevented extensive donations of land to religious houses simply because there were few landowners who could make such donations. For those who did own land, a strong earl was a role model; if he was not personally enthusiastic about monasticism, his neighbors would follow his example.[6] The influence of a strong aristocracy can also be seen in the numbers of new recruitment. As R. B. Dobson notes, late medieval noblemen were unenthusiastic about choosing a monastic career for their sons, and monks predominantly came from the 'middle classes.'[7]

Third, religious houses had certain exclusive rights regarding people and the land. Religious themselves came under canon law in the Church Courts (which were often thought to favor them). In addition, there were some particular privileges of the Church that extended to monasteries as well; perhaps the most noteworthy of these was the right of sanctuary, which, simply put, meant that anyone claiming protection inside a church or a monastery would be granted sanctuary for a certain period of time. This right was not always easy for the Church itself, but it offered protection and as such it was valued. There were occasions, however, when this privilege caused great irritation; there are cases in the royal courts that illuminate the difficulties relating to this particular right.[8] For example, the Abbot of Tavistock was accused of giving sanctuary to a man accused of murder. The brother of the victim, named John Nelar, sued the Abbot for harboring a murderer for two weeks.[9] During these difficult times rights otherwise accepted, such as this, could turn against the religious. This particular right was under scrutiny by Thomas Cromwell, who wanted to abolish it and re-establish it in a secular form.[10]

The fourth aspect of monastic power is the question of obedience inside a religious house. As we have seen, religious who followed the rule to the letter could not even consider being disobedient. A good monk or nun should concentrate on obeying and respecting the power of the head of the house as Chester's own St Werburga had done:

> Lowly submyttynge her / vnder subieccyon
> Vertu to encrese / myndynge moost relygyon
> She refused yet more her owne proper wyll
> Put all to her abbesse / her order to fulfyll[11]

Werburga's humble virtue is especially remarkable because of her royal background. She shows her submission by helping and serving others. Even when she is abbess she uses her power not authoritatively but humbly (she is rather a minister than a mistress), subsequently gaining everyone's respect and love by her example:

> She was a mynyster / rather than a maystres
> Her great preemynence / caused no presumpcyon

She was a handmayd / rather than a pryores
Seruynge her systers / with humble subieccyon[12]

Presenting St Werburga, a princess, as a handmaid is important since it denotes putting oneself in the lowest possible position. Even though being a handmaid is a common metaphor, in Werburga's case it gains strength because the princess prefers being a servant to being an abbess, and wants to serve her sisters humbly. Such humility is required of all religious heads, but at the same time they have to keep in mind that humility, too, is difficult and even dangerous; in practical terms the relationship between the abbot and the monk, between the abbess and the nun, or between the prior and a friar is always complex. Familiarity between the abbot and his monks can lead to destructive loss of authority. In a community where subjects are not in a position to complain without endangering their souls, authoritarian rule is thought to serve them best. Whatever the monk is asked to do, he has to accept it with a glad heart. This alone is true and voluntary obedience, and a demonstration of true humility.

Richard Whitford exhorted religious to flee from all honor and all positions that would put them above their brothers or sisters, because power put the soul in great peril.[13] This is one of the greatest contradictions in religious life: someone had to be elected head of the house even though this could seriously endanger the soul of the elected.

The conduct of religious heads was frequently discussed. This position was especially demanding, and failure to discipline the brothers or sisters could herald destruction. The head's duty was to guide others with the example of his or her own life: setting a good example would maintain discipline and high morals in the house, whereas a bad example would bring devastation. If an abbot

'clene put away his poyson of propriete / in the selfe', it would certainly 'nat long remayne in the subiectes / but shortely vanysshe / faynte and deye.'[14]

Absenteeism[15] by heads of religious houses was strongly disapproved of, since imitation of their exemplary behavior was so important to monks and nuns. Abbots or abbesses who traveled on business or spent time at their manors elsewhere were problematic, because during their absence they were unable to keep an alert eye on their monks or nuns. This applied to parish priests as well, and there had been repeated calls for reformation on this issue. Many of these voices were heard inside the

Church. For those who wanted to criticize the monasteries, absenteeism was a fertile ground on which to build satires and stories about abbots occupying great manors in the fashion of the nobility and expressing contempt for their more austere monastic buildings. Now these abbots became nothing but the pimps of harlots, who preferred castles to cloisters. Instead of being teachers and fathers to the brothers, the abbots were their seducers and deceivers, and the destroyers of their own houses.[16]

It was inevitable that abbots sometimes failed, but it is actually not these concrete failures that invited all the mockery and attacks from their critics. Much of the criticism was simply playing with imagery. People could write about failures in attending midnight services or about selling the lead from the roofs, turning the monasteries into mills or simply selling religious houses for secular farms. For the founders of the houses, this would mean misery since their investments in the afterlife were wasted, and the divine services already paid for forgotten.[17] Relying on monastics was regarded not only as a waste of money but also as subjecting oneself to simple deception.

The emphasis of these critical texts was so often on housekeeping that it seems that for many critics the material aspects of the cloistered household were more important than spirituality. That Thomas Cromwell shared this attitude is revealed in a letter to the Abbot of Woburn written in 1533. Here profit is given an important role: Cromwell points out that religiosity in a good abbot is manifested in his good policy, in the 'good & welthy state and condition aswell in catoll as in corne...'[18] The measure of devotion and religious excellence seems to have been the barrel. It is ironic that elsewhere we find abbots being derided for thinking in this way. This contradiction tells us about the nature of the religious debate: it was clearly unsystematic, and religious were attacked with all weaponry to hand and from any vantage point.

How did abbots themselves perceive their role? The representations of the religious in their own texts not surprisingly reveal a somewhat different approach. In the light of the evidence found in abbots' letters written during the dissolutions it seems that they valued their office highly. They thought that only a person of mature age and good manners should be elected the head of the house – successful leadership required great qualities.

The abbots' representations of the ideal monk were similarly straightforward: he obeyed the abbot, was never malicious, and diligently attended lectures and services.[19] Even though there is no direct evidence from abbesses, it can be safely assumed that they would have agreed with their male colleagues in these matters: a perfect nun presumably had the same virtues as a perfect monk. Abbots and abbesses also seemed to have understood that ideals and reality did not always match. It would have been impossible to expect all religious houses to be havens of peace at all times, but there is evidence to show that abbots and priors were deeply concerned about the reputation of their houses during any disturbances. This indicates that religious heads were well aware of the ways in which things should have been. They acknowledged that, for outsiders, the reputation of the houses appeared as the reality.

One case brought to Star Chamber clearly illustrates a superior's concern for his reputation and that of his house. Giles Huncastle, a layman, reported that his brother had been stabbed to death in the Cistercian abbey of Combermere. The prior, the monks, and the lay brothers had allegedly concealed the victim's body in a cell. According to the plaintiff, the secrecy was explained by the prior's concern for the reputation of his house. The prior, Thomas Hamond, he says, had been tormented about the reputation of his house, which already had a tarnished image, and for this reason he demanded that the murder be kept a secret. If not he feared that the house would be destroyed:

> This abbey is allredy in an euyll name for... mysrule. And therfore he wold haue this murdre kept secret and that all shal not be opynly knowyn for the Abbey shuld be vndoone foreuer.[20]

Because this is the plaintiff's account, we cannot assume that this is an accurate representation of the abbot's feelings, but what we can deduce is that the plaintiff at least found it plausible that an abbot would think like this. His account probably raised no eyebrows. It is, furthermore, possible that his account was not based on his imagination alone. Everyone understood that rumors and wagging tongues were dangerous and harmful. Having no great fortunes to protect them, smaller houses had to be especially careful; unruly houses had unruly heads – no one wanted a reputation like that.

Not all houses were perceived as failures, however. Against the popular theory of late medieval monastic decay is the apparent success of orders such as the Carthusians. They won the respect of the English people especially because of their solitary asceticism and the voluntary austerity of their life. Resisting temptation was of course admirable, and the Carthusians were admired because their solitary cells were exposed to the attacks of the devil.[21] Further proof of the success of the Carthusians was provided by the fact that there were always more applicants than there were cells to receive them.[22] The triumph of the Carthusians was for many a proof that there were abbots able to fulfill their duty as monastic heads. Two of their houses especially, London Charterhouse and Mount Grace, amassed appreciative comments over the years. Weighty words of approval could even come from an unexpected direction: Henry's propagandist Thomas Starkey (who had no reason to defend any religious houses, as it was his duty to speak against them) wrote that the monks of the Charterhouse were 'notyd of grete sanctyte.'[23]

In contrast, some orders had poor reputations. The Augustinians, for example, were often scorned because of the signs of sinful living they showed. The London Augustinians were notorious for their absenteeism from night masses: only one priest would sing, accompanied by a mere handful of children. That house was said to be completely corrupt.[24] In the end, all such failures were failures of the head of the house, who was branded as incapable of fulfilling his duty and keeping discipline in his house.

Some criticism was directed at the monastic rule as a whole.[25] The rule was sometimes said to have lost its strictness and critics alleged that people had become weaker than their ancestors. Some supporters of the old faith, such as Sir Thomas More, suggested relaxation of the rule as a remedy; they thought that a laxer rule would probably be better kept than a stricter one. People who favored keeping the rule strict attested that laxity brought nothing positive with it. Lax rule generated more evil as it would be even easier to break. Changes would only be followed by 'boldnes to offende / a quyet herte in a euyl concyence.'[26] Those who supported strict discipline thought that a harsher rule was more difficult to break. They insisted that the rules were made by holy fathers with the help of the Holy Ghost. The rule was made to be kept: Christopher St German, for example, stressed that what was wrong with the religious life was not the fault of the religious ideal or the founders of the houses. It was the people who failed.[27] For a pessimist, there are not many who could keep the rule. For an optimist, St German's message is that it is not the religious life in itself that is evil, but malpractices which should be weeded out.

Politically aware reformers made sure that monastic obedience to the monarch became an issue in the 1530s. They repeatedly alleged that not only did the religious refuse to obey the king or God, but they would bring destruction on all by fleeing to the pope for help. Laymen, it was argued in the anti-papal camp, were better people, since they knew whom to obey and where their loyalties were best placed. Unlike monks, they would help their king defend the country.[28] The strongest critique aimed at religious insisted that disobedience was altogether the making of bishops, monks, and friars, who 'teacheth the people to disobeye their heedes and gouerners / and moueth them to ryse agenst their princis and to make all comen and to make havoke of other mens goodes.'[29] This was a formidable accusation and speaks of attitudes toward religious. Their power was feared. Had the defenders of the monks wanted to turn this in their favor, they could have tried to show that such opinions demonstrated the importance of the religious: if they had the power to destroy the whole world, they had to be very special. They did not, however, use this argument. Thomas More, who furiously opposed these views, thought that the heretics were mad to suggest that the clergy encouraged people to rebel against the king. The very idea was preposterous; the clergy was not so stupid that they would bite the hand that fed them.[30] The point of the reformers, however, was that the obedience of the religious was a relative issue. Because they could choose their own leaders, they need not obey anyone outside their house and could serve their head in the way they themselves thought best. It was a circle of disobedience and utterly evil. If obedience was twisted for selfish purposes, it certainly was not difficult to maintain, since religious were required to please only themselves in the end. Even the Ten Commandments could be broken, since the religious honored no one, not even their parents. Rather, their mothers and fathers honored *them*.[31] It is interesting, however, that the reformers' view here presents religious as obedient; but their obedience was merely aimed in the wrong direction. A closer look at the lampoon *Reade me frynde and be not wrothe* throws light on the issues relevant here and continues to support the paradox: in being obedient, religious actually are disobedient. First, *Reade me frynde and be not wrothe* concedes that religious are obedient, but problems follow because of the nature of their obedience:

> I will not denye they do obeye,
> Vnto the ruler of their abbeye,
>   A carle of their owne chosynge.
> Yet it is in superstycyousnes,
> Without any profytablenes.

The obedience of the religious is directed to an abbot whom they choose themselves. This kind of inverted obedience was of no value. After organizing their lives so that no genuine obedience is needed, the monks can concentrate on pleasing themselves. The author seems to feel that people are under the power of religious monsters. Furthermore, good people have to serve and obey the king in all things whilst religious care nothing for their king, and turn to the pope for help:

> To serue the Kynge in warre and peace,
>   The defence of the realme assystynge
> Where as the relygyous sectes,
> Vnto no lawes are subiectes,
>   Obeyinge neyther God nor kynge.
> Yf the Kynge wyll their seruyce vse
> Forth with they laye for an exuse,
>   That they must do Goddes busynes:
> And yf in it they be founde negligent,
> They saye the Kynge is inpediment,
>   Because they must do hym serues.
> And yet the Kynge shall them compell,
> Then obstynately they do rebell,
>   Fleynge to the Popes mayntenaunce:
> Of whome they obtayne exempcions,
> From all the iurisdiccions,
>   Of temporall gouernaunce.[32]

This extract gives us a glimpse of the complex relationships between monks, abbots, the pope, and the king. It puts into focus two of the most conspicuous issues of monastic obedience that interested the evangelicals. The first was the role and election of the head of the houses. His autonomy particularly disturbed the critics, as they claimed this placed him beyond the reach of the king. The second was the relationship between the monks and the king. It was clearly a formidable issue. Some feared that monks would soon rise up against Henry in defense of the papacy, and there seems to have been hard evidence of sedition among religious.

This evidence started to emerge especially in the months before the royal divorce. Queen Katherine seems to have found support in those religious orders that were known for strict observance of the rule: the Observant Franciscans, the Brigittines, and the Carthusians. These were the orders that had gained most respect, and if *they* ignited conspiracy, it was only reasonable to fear that they could win people to their side. This fear was not unfounded, but it is a different matter completely (and beyond the scope of this study) if there were any true seditious intentions in these orders. Nonetheless, the Observant Franciscans especially on the queen's side could have been harmful to Henry, because of their influence in the

Tudor court. For decades the royal confessors had been Observants, which gave them a position of immense religious and even political status.[33] The danger they presented came from two directions: from inside the court, through their close contacts with the pious queen, and from outside, even from the continent, where the Observants had close contacts. By 1533 at least, Thomas Cromwell was convinced that the friars were conspiring against his master. Cromwell interrogated some of the Observants, but little emerged other than a note that one of them, Hugh Payne, was 'a veray sedycious person.'[34]

Cromwell's interrogations did not lift the cloud of suspicion that hung over the Observants or the Carthusians. In 1534 when the trouble with the Act of Supremacy began, members of these orders found supporting the king's cause impossible, and inevitably found themselves in a perilous situation. The Greenwich Observants and the Carthusians of London Charterhouse were under oppression. The case of Friar Forest (an Observant from Greenwich) is well known; he refused to accept the royal supremacy, and in a letter to the queen written in 1534 revealed that he was prepared to die because he could not take the king's side. He even sent his rosary to her as a farewell gift.[35] Subsequently, however, Forest changed his mind, took the oath of supremacy and was thus able to avoid the death of a traitor. This was only short-lived; in 1538 he was burned as a relapsed heretic and traitor who still regarded the pope as the supreme head of the Church. A poem nailed at the site said:

> And Forest the frier,
> That obstinate lyer,
> That wilfully shalbe dead,
> In his contumacie
> The Gospell doth deny,
> The Kyng to be supreme head.[36]

Friar Forest's execution was a special occasion, at which Bishop Latimer preached to a large crowd consisting of dignitaries. They all watched the friar burn with a statue of a Welsh saint, St Derfel. Forest's immolation with the statue was carefully orchestrated since St Derfel's statue was not just any image: the saint was renowned for saving souls from Hell. Margaret Aston has attested that there had been a prophecy attached to St Derfel: he would one day set a forest on fire.[37] In a recent article Peter Marshall suggests, however, that this prophecy was a later invention and thus bears no contemporary significance.[38] Because there is no evidence to support the theory of the early existence of the prophecy,

Forest's end is somewhat stripped of its romantic symbolism. Neverthe-less, his death speaks of the fear of the most powerful religious. It says: they had standing in society, they had religious power over people, and they could be dangerous.

Forest's case is just one example. There had been several occasions of disobedience toward the king in religious houses before Forest met his end. These cases disclose pieces of our puzzle and offer us rare infor-mation about how the monastics themselves felt about their lives and their vow of obedience. As a group, it was the Carthusians (especially in London) who were, in the end, the most decisively united against the king. According to the king's visitors, they were 'wonderfully addicte to theire old Mumpsimus,'[39] popish nonsense, horribly obstinate, and determined not to accept the king as the supreme head of the Church. The fate of the Carthusian priors of London, Beauvale, and Axholm, on May 4, 1535, was famous in its time. As the romantic story told by Nicholas Harpsfield depicts, Thomas More – himself a prisoner in the Tower – watched the priors being taken to their execution. More had envied their fate because those 'blessed fathers be now as cherefullie going to theire deaths as bridegroomes to their marriage.'[40] As religious were traditionally thought to unite with God in marriage after death, this was a direct allusion to their special role in God's eyes.

The very same day the London Carthusians were sent anti-papal books. This was intended to change their obstinate minds and probably also to keep the monks occupied otherwise than in propagandizing the priors' deaths as martyrdom in the pope's cause. The monks did not yield. They were determined to suffer rather than to switch alle-giance, and the books were returned.[41] As if Cromwell was testing the Carthusians' steadfastness, three more brothers of the Charterhouse were taken prisoner and incarcerated in the Marshalsea. They were chained to a wall for two weeks, but remained steadfast and refused to accept Henry as their head. They too were executed.[42] Oppression of the London Charterhouse now became harsh. What were perceived as popish books were confiscated, some brothers were sent to other houses, and the remaining brothers had to accept six lay guardians in their house. In May 1537, ten still obstinate monks were taken to the notori-ous Newgate prison where they were left to die, one by one, during the summer.[43] Yet, one of the monks survived, and was transferred else-where. He was, however, executed in 1540 for remaining faithful to the pope.[44] The fate of the Carthusians gained publicity in London,[45] but leaving the monks to die rather than execute them was certainly strategically well thought out: the king had no need for any more martyrs, and when the prisoners were 'dispeched by thand of God,'[46] that is died from natural causes from the wretched conditions they were thrown into, they were seen as much less dangerous than martyrs to the papal order.

This harsh treatment of the Carthusians seems less irrational if it is remembered that at the same time Henry was dealing with several uprisings in the North of England. Henry had always had problems with uprisings, and must have feared that any one of them could become a threat to his throne. Any serious unrest in London could easily have turned into a catastrophe for the king. By 1537, several uprisings had already proved that even religious – who were not legally allowed to carry weapons[47] – were ready to stand against the king's officers. In the revolts, religious motivation, as the present-day scholarly consensus[48] claims, was a significant factor, although not the only reason to rise up. After the suppression of several smaller religious houses, it is no surprise that some monasteries in the North took part or were forced to become involved in the revolts. The Pilgrimage of Grace was especially pronounced in its religiosity: the rebels used the symbols of pilgrims and declared that they wanted to express the passion of Christ in their actions and carried a flag that depicted the five wounds of Christ. They also swore an oath to restore the religious houses that had already been dissolved. The interrogations of the leader of the Pilgrimage, Robert Aske, prove that in his mind the uprising was profoundly religious; on several occasions he mentioned that the uprising had flared up because of the dissolution of the monasteries. The suppression of several houses in the North gave ground for strong rumors about further religious changes that would touch everyday devotion in a more profound way than the suppressions ever could. It was rumored that jewels and vessels from all the churches and chapels would be confiscated and replaced by cheaper materials. Furthermore, parish churches at a distance of less than five miles to their neighbors would be dissolved, and a tribute would have to be paid to the king at every christening, wedding, or funeral. All this gave ample reason for a rebellion.[49] In the sparsely populated North, religious houses had more important practical functions than in the more populous South.

The Walsingham conspiracy posed the most significant threat of rebellion in the South of England. In early 1537, a group of Norfolk peasants planned a rebellion against the suppression of monasteries and against the gentry in the region, but they did not arm themselves before being betrayed in April. The conspirators included the sub-prior of Walsingham and three Carmelites from Burnham Norton; two were executed and one freed after a short imprisonment.[50] The aim of this conspiracy as a religious uprising is clear: the dissolutions had to be stopped and the status quo restored.

During the uprisings, Henry's worst nightmares certainly seemed to be coming true. The obedience of the religious appeared, after all, to be directed toward Rome. Many religious were involved in the uprisings, and in the aftermath several were executed. Trying to talk their way out of trouble, some of the religious who were accused of conspiracy insisted that they had been forced to join the rebels. One of them was the Abbot of Whalley, who insisted that the rebels had threatened to burn the Abbey crops if he did not join them. The Abbot of Jervaux, for his part, claimed that he had tried to escape the rebels, but had been caught and forced to join under the threat of burning down his monastery.[51] Regardless of the examples of these reluctant rebels, it seems that the Northerners were genuinely concerned at the loss of the religious houses, which they saw as irreplaceable. They were adamant that this kind of disobedience to the king was not against God's will, as they thought they were right and were doing God's will. Despite the fact that Robert Aske had ordered the religious not to join the rebels, many of them found the religious life worth risking their lives for. Among them were the monks of Sawley, who returned to their suppressed house, thus openly defying the king's will. This resulted in an order issued by the king to execute the disobedient abbot and monks immediately.

The king's suspicion that religious were involved in the uprisings was further demonstrated by his conviction that the Lancashire revolt was solely the making of monks.[52] The executed rebels of the Lincolnshire revolt, moreover, included four monks from Barlings Abbey and six from Bardney. The Abbot of Kirkstead and three of his monks met a similar fate.[53] Monasteries, regardless of the Act of Dispensations in 1534, were still tied to the pope in people's minds, and Henry now had more reason than ever for thinking that the monasteries were leading the popish campaign against him.[54] Henry's views were corroborated by reports from his commissioners. Sir William Fayrfax, for example, warned the king that monks were able to persuade people to take their side: 'these two sorts hath no small number in their favors, arguing and speaking.'[55] Thomas Starkey, who was close to the king at that time, confirms that there indeed was great controversy about the dissolutions.[56]

Religious houses can hardly be regarded as organized military centers of resistance, but they were influential. One of the most effective of their methods was to distribute prophecies against the king's cause; the case of Elizabeth Barton, the Nun of Kent, was of recent memory. Some examples can illustrate their nature. Two monks from Furness, John Harington and John Broughton, publicized a prophecy which declared that the pope would regain his power in England in three years' time, and Robert Dalivyl prophecied that the king would not live after St John the Baptist's Day in 1538 unless he repented and mended his ways immediately. Another monk reported that he had seen an angel that exhorted the English to rebel against their oppressors.[57] To a king who had always feared rebellion, these monks presented a true threat. Prophecies were often taken very seriously.[58]

There were several monks who died defending their house during the dissolutions, besides those executed during the rebellions. Although they had submitted to swearing the oath of supremacy (often with mental reservations), they could not accept that the monastic tradition would be suppressed after its thousand-year history. Even though they were a small minority, their obstinacy in the face of death shows the ideals of the religious: they were prepared to die for the right cause, and monasticism was a worthy one. Also, the cloak of a martyr for many, perhaps surprisingly, was a welcome one. For many abbots and abbesses it seemed impossible to remain obedient to the king when it was their turn to give up their houses. Here they had to defer to the pope.

In order to avoid dissolutions, several of religious houses tried to bribe the king or Thomas Cromwell, who was the main executor of the dissolutions. A strong desire to keep his monastery alive seems to have been the motive of, for example, the Abbot of Croyland, who sent Thomas Cromwell a fish. The Abbot of Leicester sent him some other foodstuffs.[59] Richard Whiting, the Abbot of Glastonbury, offered money, but when the commissioners found out that he had (allegedly) concealed some of the valuables and money in the grounds and the vaults and walls of his house, he was accused of robbing his monastery church and executed.[60] The Abbot of Reading also offered money in exchange for his house.[61] The Abbots of Glastonbury and Reading belong to the group known as 'the three abbots' whose executions marked the culmination of the rather paranoid and probably exaggerated attitude toward the might of the monasteries. Together with the Abbot of Colchester, they were executed in 1539 as a warning to others who might have wanted to oppose the dissolution of the remaining large houses. The three abbots were branded as ruthless, deceitful, and treacherous: 'treason that hath be of long tyme hydde in huckernucker among as they were called religious men.'[62] If obedience to the king was not automatic and immediate, the outcome was terrifying.

Many heads of religious houses thought that when it came to their own house, the requirement to be obedient to the king did not apply. The reasons for this exclusion and for not surrendering the monasteries were manifold: Prior Hord of Henton refused because he felt that his house was not his to give. The house belonged to God, not to the abbot or to the monks – nor to the king.[63] Prior Hord's reasoning was sound, but useless. The king's claim was well founded as well, and many submitted with little resistance, in the belief that as the supreme head the king was legally able to make decisions about church property. Moreover, his personal word was believed: Cromwell's commissioners reported on several abbots or abbesses who refused to give up their house to the commissioners, but said they would certainly do so if Henry asked them in person.[64] In the end, the main question in the dissolutions was about obedience. Was it due to a king who was destroying their way of life? When faced with the plain facts, most people and most religious decided that it was. They simply had no weapons against the royal demands.

The North was especially loyal to the monasteries. Claire Cross has studied Yorkshire wills and has shown that many religious still lived together after the dissolution, and that the vitality of monasticism 'may have been little less feeble' than has been previously supposed.[65] M. A. Clark's study of Cumbria corroborates this view.[66] This evidence of determination to continue the religious life suggests that surrendering the religious houses to the king did not necessarily mean rejection of the religious life. Obedience counted; there was no other choice for the religious people but to submit – or die.

The fact that 53 religious houses already shortlisted for confiscation during the dissolutions of 1536–7 were granted permission to continue shows that many religious were anxious not to give up their religious life, for such permission was not granted without the activity of the religious themselves. It must also be noted that of the 103 small nunneries that were listed for dissolution in 1536, 28 were allowed to continue as religious houses. This clearly indicates the active commitment by the nuns to their vocation.[67]

Eagerness to continue the religious life varied, however. When offered a chance during the visitations and later during the dissolutions, many willingly gave up their religious habits. There was some variation between the orders, but as Clark notes, even some Benedictines, who were often considered the most corrupt of the orders, were willing to offer resistance. His example is St Albans whose monks 'continued to live in close fellowship' after the dissolution in 1539.[68] David Knowles has estimated that almost half of the Augustinians renounced, but only one tenth of the Cistercians.[69] It is somewhat ironic that the commissioners, who often seem to have been evangelical, found the monks' willingness to obey Cromwell despicable. Even obedience to the king could thus be turned against the religious. Richard Layton, for example, wrote to Cromwell that many monks were tired of everything the habit stood for: 'suche religion and faynede sanctitie gode save me fro!' he exclaimed, and concluded that this had to be a sign of the monks' hypocrisy.[70] The everyday reality of the dissolution period was that nothing seemed to satisfy the commissioners.

We have now considered the evidence, some of obedience to the king and some of resistance, but what can be made out of it? It must be stated that in the end the majority of the English people and even the great majority of the religious houses remained obedient to the king. Before they submitted, there certainly were murmurings; and most people were disturbed when rumors circulated about new taxes and the abolition of parish churches or chantries. No doubt the dissolutions caused at least some degree of confusion among the people. As we have seen, there was resistance to the actions taken by the king, but not enough to stop the suppressions. Most religious seem to have wanted to continue their way of life, but they were faced with a contradiction: they believed their way of life was worth defending, but standing up to a king who executed all his enemies was difficult and presumed a willingness to martyrdom. Not all were prepared to die for their houses; many were, though. Whether the monks and nuns decided to obey the king or the pope, they ended up being disobedient to one or the other. It was an insoluble dilemma. On the one hand, most religious agreed to swear obedience to the king and accept his supremacy over the Church; on the other, they were reluctant to follow through the changes at hand when the target of the reformers became their way of life. As we have seen, when individual houses were threatened, some monks even took up arms and were executed as traitors. Some sixty monks were executed because of their disobedience.[71]

The vow of obedience was still important to religious. The fact that so many remained faithful to their houses and their religious life, and even (though much less often) to the pope, meant that the fear that gripped the king and his servants of a widespread war in which religious would join the pope against the king, was in some minds close to realization.

I have attempted to show that obedience was not an empty vow in the religious life. Religious were depicted as disobedient in two ways: either as disobedient mules, who had only their own gain to think about, or as obedient only to those whom they wanted to please, for example their abbots or the pope. If this is approached from the point of view of the religious themselves, or anyone who supported the old religion, their disobedience immediately turns into obedience. What the critics saw as treason against the king could, from another angle, be seen as righteous obedience based on monastic vows and the very basics of the Catholic religion. It could be seen as loyalty and determination to continue living a life in which they had sworn to live until death. In this way the criticism of monastic disobedience provides further evidence that obedience was a major feature in the representation of religious.

Having said that, we should note that the concept of obedience supported by the evangelicals hardly differed from the traditional idea of obedience: they merely argued that obedience was due to the king rather than to the pope, but as we have seen, this was not new. It was easy to justify this claim for example by reference to the scriptures. Obedience remained as important as it had always been, and even its expressions were changed little by the reformers. As obedience by religious was curiously acceptable even to the reformers if they unquestioningly obeyed the king, it seems that the new order was building very strongly upon the old one. Perhaps this was their strategy all along.

Part II
Paupertas

4
The Problem of Having Nothing and Owning Land

O mercyfull god /
what money is able to bryng man to heuen?
Or where myght we fynde a marchaunt
in this worlde able to make this banke?
William Bonde: *Pilgrimage of Perfection*,
sig. H5v–H6

Obedience to the rule required a life led according to the vow of *paupertas*, voluntary poverty. This contradicts the popular notion of late medieval monasticism: rich houses and wealthy abbots. Part II discusses the idea of poverty and contrasts the idea with the apparent wealth of the richest houses. This chapter focuses on landownership and chapter 5 on other property, the 'riches'.

In short, the economic situation of the religious houses was as follows: There were approximately 880 religious houses and hospitals in England at the beginning of the 1530s, and approximately 11,000 male religious and 1,600 nuns lived in them. More than half of the houses received an annual income of more than £100, which in most cases left them reasonably well off. It must be noted, however, that even though many houses were wealthy, only 4 per cent had an annual income of more than £1,000, which is usually considered the watershed between the medium and the grand.[1] In local terms, the large monasteries were important centers of economic and social activity. In Reading, for example, the Abbey fair was extremely important to the local economy.[2] The statistics compiled by R. B. Smith show the economic significance of monasteries: in the West Riding of Yorkshire the annual income of the religious houses was £4,640, quite a sum compared to the Crown's income from the same shire, which was only £2,080, and this sum already includes the income from the mighty Duchy of Lancaster. In comparison, it can be noted that the nobility in the area had a combined annual income of £1,995, which was thus less than half the income of the religious houses.[3] These figures tell a story of a world in which religious houses could not be ignored in economic matters, but also of a monastic world which could not always boast of riches.

As landowners, monasteries seem to have been flexible. J. N. Hare suggests that they had been able to answer the demands of changing times by investing. This probably helped them through the disastrous years following the Black Death.[4] Not everybody agrees with Hare, however. Maurice Keen, for example, portrays the religious houses in a much darker light, and the Black Death as especially destructive to them. Keen maintains that the recruitment of novices, for instance, was difficult after such a severe decline in population.[5] Certainly, like all landowners who were financially dependent on the land, religious houses suffered from the rising cost of labor and collapsing rents. Some houses became badly indebted and took a long time to recover, but most houses had recovered by the end of the fifteenth century.[6] In terms of recruitment, religious orders were thriving in the later half of the fifteenth century.

Land and other riches accumulated to the religious houses through donations from pious supporters. The contemplative orders regarded landownership as necessary for their independence and autonomy from the secular power; property, lordship, and power were 'embraced . . . as the means by which the church could free itself from domination by laymen and their concomitant worldliness.'[7] What the critics of the Church interpreted as mundane greed, the Church saw as a holy pursuit for the kingdom of God, which submission to lay rule could have endangered. Viewing its role in such a way made the monastic order a significant part of the feudal system, remains of which still existed at the end of the fifteenth century. There were 400 estates with bondsmen in England at that time. Of these not more than 55 were owned by the Church and only a small number of those 55 by religious houses; these numbers offer no justification to blame monasteries for not freeing their bondsmen, but they do show that monasteries had a role to play in feudalism.[8] Some religious houses were grand institutions with grand needs.

In contrast, there were houses that could not boast of great wealth and grandeur.[9] Some of them were so poor that even Thomas Cromwell's hardened commissioners seemed genuinely shocked by what they saw. Houses such as the Shrewsbury Augustinians were reported to be in a terrible state. There was nothing there to enable decent living: no beds, no bread, nothing even to drink.[10] Not surprisingly, in most cases the poorest houses were friars' convents. The friars at times suffered from severe poverty, since their rule forbade them to own land or handle money. The Franciscans of Bridgenorth provide another good example of the poorest houses. The friars were willing to surrender their house; they were simply exhausted from utter destitution because people were so parsimonious with their alms.[11]

Originally, of course, poverty and suffering in meager conditions like those mentioned above were the monastic ideal. Thirst and hunger were thought to elevate the mind, and riches were despised. In the course of time, however, monastic poverty was transformed into a more symbolic deprivation rather than actually experiencing the rumbling of empty stomachs. By the sixteenth century, not even the poverty of the friars remained absolute. Notwithstanding their ideals, they too had accepted generous gifts and property for their convents. This was in contradiction to the original ideal of the *vita apostolica*, in which Christians gave up all their possessions to the poor and trusted that the Holy Spirit would provide them with essentials.[12] Their accumulated wealth had been seriously criticized by the medieval monastic reformers, who wished to restore the principle of ascetic poverty to the religious life. Few praised the rate at which riches piled up in the monastic barns, cellars, and granaries, but many thought that if no love of riches was involved, and the property was used well for hospitality and other acts of charity, ownership could not be completely evil.[13] Many felt that a relatively stable income for the houses from land and other sources ensured that their inhabitants at least would not die of starvation. Moreover, because religious provided the livelihood of a great number of lay people, it was not only the religious themselves who were supported in this way. Helping the religious would also help those who were dependent on them, especially the poor.

In the case of individual religious, however, the ideal of poverty was total, at least in theory. A monk, nun, canon, or friar was not allowed to own anything; the community would provide them with all they needed. Voluntary poverty was seen as an effective remedy for the sin of avarice, and therefore an effective defense against the temptations of the world. Poverty literally liberated one from the world. Because the monk owned nothing, not even his habit, he was completely free, 'voyd of wordly cares,' as Richard Whitford put it, and was rewarded with a peaceful mind. But this was, as much as anything in life, conditional. Living in utter scarcity was meritorious and laudable only if one engaged oneself in the service of God, in prayer, meditation, contemplation, and spiritual exercise.[14] This was not regarded as too demanding, since God helped those who trusted him enough to give up everything else:

> Thus God oft helpeth them, that in him haue trust,
> When worldly ryches, men leaueth in the dust.[15]

The term 'voluntary poverty' is curious. It was exclusive and elitist, applying only to those whose origins were relatively wealthy, for how could a pauper become voluntarily poor if he had nothing in the world to leave behind? The idea of voluntary poverty indicated that not all kinds of poverty were meritorious for the soul.[16] Therefore, not even poverty was what we would call democratic. The ideal monastic pauper therefore came from a prosperous background, since only he or she had something to lose once the habit was taken. *Paupertas Christi* was an acquired, not a necessary, state.[17] As ever the princess St Werburga served as an ideal example: she if anyone had a world to give up and was perfectly willing to do so. Her resolution to swap the royal diadem for the simple veil of a nun was described by the translator of her biographer, Henry Bradshaw, as a wonderful deed celebrated in heaven.[18]

We have already discussed how the monastic rule was bent to allow religious houses to accept land donations; similar adjustments were also made in regard to the personal property of religious. This is quite surprising, but in practical terms, the nature of late medieval monastic poverty was somewhat different from the early ideals even in this respect. Nuns could sometimes take their personal belongings to a nunnery. There is some evidence that possessing things was regarded as acceptable if the prior, abbot, or abbess gave consent; ownership became sinful only if possessions were kept in secret.[19] Private ownership in religious houses ensured that especially the wealthiest monks or nuns did not have to compromise the standard of living they had been used to at home. This was according to the all-encompassing nature of the social hierarchy in society. Customs such as this were not, however, accepted by all, and some orders found this practice incomprehensibly lax. The Brigittine Richard Whitford was appalled at such practices as paying wages to religious. He wrote in horror of how monks and nuns could spend their allowance on whatever they pleased: shopping, lending, or even gambling.[20] Whitford's text expresses shock; he was convinced that he was not only repeating a malicious rumor but that there were houses that had succumbed to such horrors. It cannot be asserted that this kind of relaxation of the rule was common, but clearly malpractices did occur and aroused strong disapproval. For those who valued the traditional ideals, personal possessions in monasteries were a gross transgression of the rule.

If the accumulation of wealth in religious houses aroused criticism within monastic circles, outside the religious orders it did so even more. Complaints about monks' greed were not rare in the Middle Ages. Economic issues such as taxation were close to the hearts of people, and not everyone ungrudgingly accepted their economic obligations to the Church – tithes and other taxes were burdensome at times. Spiteful tongues sometimes wagged and said that people were brought to the verge of destruction in their attempts to satisfy the needs of the monastics. Similar complaints could be heard in the sixteenth century as well. Reformers said that the monks had acquired the best land in the country by deceiving the dying and robbing their heirs of their inheritance.[21] Riches brought corruption and decay, and superfluity generated greed.[22] The extract below talks about a medieval ploughman's resentment at being expected to donate to the different religious orders while simultaneously trying to keep his family alive. The friars demand money, wheat, barley, cheese or meat:

> The come the gray ffreres and make thew mone
> And call for money oure soulus to save
> The come the white ffreres and begin to grove
> Whete or barley they woll fayne have
> Then cometh the freres Augustynes & begynneth to crave
> Corne or chese for they have not inough
> The cometh the blak freres which wolde fayne have
> I pray to god spede wele the plough.[23]

To the ploughman it is a personal punishment to have to sustain the religious orders. As can be seen, it is not only the contemplative orders that are blamed for sucking the husbandmen's blood: the mendicant friars, even the holiest, are guilty of similar greed. The clergy at large were often accused by Catholics and evangelicals alike of caring about worldly riches more than about the souls that they were supposed to save. Many seemed to seek only worldly glory and honor and to forget the glories of heaven in their pursuit of the respect of their peers. As Christopher St German put it, the clergy served God for worldly praise only.[24]

The wealth of the clergy, their extravagant lifestyle and the way property lured them into worldly activities were the main themes discussed by Simon Fish, a harsh, perhaps Lollard-minded critic of the

Church.[25] One of the Lollard complaints was of the excessive wealth of the clergy; Fish himself claimed that the clerics had stolen one third of the arable land in England. He felt that this outrageous injustice should be redressed by confiscating the property of the Church and giving it to those who truly needed it.[26] Fish's allegations provoked Thomas More into a fierce defense of the clergy. Impatiently, he fulminated that people no longer donated enough property to the Church to give cause for any such critique, and furthermore, he stressed, even if people *had* continued to be as religious, virtuous, faithful, and generous as they had been in earlier times, there still would have been no grounds to fear that all the wealth of the nation would shift to the clergy. More thought Simon Fish was a despicable liar and was especially infuriated by Fish's claim that the greedy friars oppressed people into giving them money by threatening to denounce them as heretics if they refused. More fought back by arguing that in no heresy trial had a friar been accused of such oppression,[27] and More if anyone should have known as he was a persecutor of heretics. Had this occurred, someone would certainly have spoken his mind about it to More was the message sent to Fish.

Besides Simon Fish, there were others who expressed themselves strongly against religious. William Tyndale wrote that the 'countenance and quality' of present-day monks were far inferior to those of the first monks, which had been impressive: he praised their abstinence and their readiness to earn their living by their own labor. This praise immediately raised the question: what had happened to all this virtue? The answer was not the evil nature of religious, as perhaps could have been expected, but their very excellence. Everything had been turned upside down because people were so taken with their goodness that they wanted to make donations, and this opened the pit to hell. Wealth poured in and made the monks greedy. Instead of distributing the wealth to the poor, they wanted more. Tyndale further declared that when the decay of monasticism had become apparent, the various orders of friars were created in order to save whatever was left of the credibility of the Church in general and the reputation of the religious life in particular. For a time the friars had managed to deceive the people, but soon enough their greed emerged as well.[28]

The past was a handy propaganda tool, presented as a golden age when even the poor had sufficient, when there was an abundance of food and everyone lived in freedom. The clergy had been modest, merciful, satisfied with little, and their hearts were pure, setting an excellent example for anyone who wished to follow Christ's path. Yet all this had been destroyed when the clerics got their hands on property.[29] Friars were hardly better:

> They robbed the worlde vtterly,
> Causynge it with extreme beggery,
> In greate ruyne to be decayde.[30]

The outcome of the clergy's immense power was depicted in disillusioned tones. For evangelicals it seemed almost impossible to find anything good in the present situation.

The past excellence of religious institutions could not be restored by the Church itself. The distant past portrayed as a happy haven was imagery used in Henry VIII's propaganda as well. Richard Morison thought that the clergy had lost immensely because they had started gathering riches to their hearts and hands.[31] When he was writing, the dissolutions of the smaller religious houses had already begun, and Morison expressed genuine surprise at reactions to this; he was amazed to hear people complaining about the dissolutions, as he clearly expected that people would be in favor of the confiscation of church property. Morison asked bitterly if people had already forgotten their complaints about monastic wealth and its misuse; had they forgotten that monastic property was to be divided so that it benefited everyone?[32] Morison's reaction shows that not everyone perceived monastic wealth as a justification for the dissolutions. If we focus on the views presented by monasticism's critics, we are led to a very different and misleading conclusion. This is what 'the Protestant' scholarly tradition did for centuries after John Foxe in order to justify 'the Reformation.'

To correct this distorted picture, let us now consider other evidence speaking for religious, in addition to Morison's indirect implication. We have already briefly considered Thomas More's ideas when discussing Simon Fish's allegations, but I will now return to his ideas about religious. More was extremely critical of the Lutherans, whom he branded as foolish and illogical. When he discussed the errors of heretics, More's tone was often sarcastic, and when he wrote about the complaints of his opponents about the property of monks and the greediness of friars, he revealed a certain degree of cynicism. More wondered why the Lutherans never bothered to complain about nuns' possessions, who, after all, held vast estates and were by no means on the brink of starvation. More answers this question himself: Lutherans did not bother with nuns simply because they considered nuns to be irrelevant, since as women, nuns had no authority in society.[33] More's point is clear. The whole polemic about monastic possessions and the wealth of the Church was about power and power only. The question was, who could exercise power over whom? If power was important in the debate – and there is no reason to doubt More – it is no wonder that nuns are not more often the focus of religious controversy. Nuns were fewer in number than male religious, on average their houses were not very grand, and their spiritual influence was certainly greatest in the sphere of other women. Cloistered away, without significant public spiritual authority, nuns were invisible and ignored. But there was one formidable exception to the rule: Elizabeth Barton, whose voice as a female religious prophet threatened to shatter the whole Henrician construction of power.[34]

Above, we have discussed the representations of monastic ownership in printed sources, but we should now take a look at court cases and discuss the perspective they offer on the question of monastic wealth. Several cases involving religious survive in the records.[35] The complaints are usually quite brief, but they describe the criminal offenses at some length and are particularly revealing about attitudes toward abbots. The Abbot of Bindon, for example, is described as *influential and wealthy* in a case brought to Star Chamber by an anonymous plaintiff. The plaintiff deplores his own poverty and misery compared to the abbot, and suggests that the abbot could easily replace some sheep the plaintiff has lost to the abbot in a feud.[36] *Criminal deception* was in the mind of Philip Brayne when he accused the Abbot of Buckfast of, as it were, kidnapping Brayne's father-in-law to the monastery to die. His father-in-law had died just three days later, and the abbot had taken everything the man had owned: twenty sheep, two horses, and seven bulls. The abbot's only motive, Brayne claimed, was the unlawfull seizure of the old man's property.[37] The probable and logical explanation is that there had been an agreement between the old man and the monastery that he would move to the monastery in his old age as a corrodian. This was not unusual; the animals would then have been spoken for as payment for his upkeep. Sudden illness made the abbot worry about the outcome: his house would gain nothing if the man died before he moved in. From the perspective of the dead man's daughter and son-in-law, the abbot's motive had been not to ease the old man's soul's journey to heaven, but to acquire the property by unlawful means: the abbot's greed had overtaken him.

A case against the Prior of Leominster speaks of extortion and oppression. John Galet accused the prior and a group of his men of physical assault, which had been so violent that Galet had feared for his life. He calls the prior *evil*, *mean*, and *disobedient* to the king and his council.[38]

The motive for the attack remains somewhat obscure, but it seems once again to have been a feud over land.

In another case, the Prior of Malton is described as *murderous*, having allegedly masterminded an attack on William Perchey, his family, and guests. Canon John Jackson, John Gaton, and Thomas Redhead had attacked the group (which included Perchey's young daughter) while they were taking a walk. Because the men had been armed with daggers, poles, and dungforks, Perchey insisted that they had intended to murder them all. Luckily, thanks to Perchey's successful use of the sword, everyone had been able to escape to the manor and save their lives. The reason given for this murderous spree was unpaid debts.[39]

Curiously enough, these cases seem not to reveal any malevolence against religious in general. Even the most hair-raising case I have come up with shows no such general resentment. This case is so fantastic that it must be at least partly true, and I cannot resist the temptation to recount the tale here.

William Lokewode and his wife Agnes brought to Star Chamber a case against Robert, the Abbot of Sawtry, whom they accused of theft and illegal imprisonment. The account is as follow. First, Abbot Robert arrested Mistress Agnes and took her to the monastery cattle shed, and kept her there for three hours. After two days the abbot repeated this, but now arrested Agnes's parents as well, and the three were kept prisoner for nine weeks; what is more, they had to pay for their upkeep during those weeks with Grandmother Johane's ring. Meanwhile, the key to Johane's coffer was stolen and the contents taken (it is not mentioned who the culprit was, but it is indicated that the abbot was to blame). In addition to this ordeal, the abbot arrested and stripped naked Johane's maid in the chapterhouse, an act of impropriety which the plaintiffs especially disapproved of in a religious. The events do not end there, however; the Lokewodes' five-year-old son was taken next, whipped, and questioned about Johane's property. Furthermore, the abbot struck the boy's 80-year-old grandfather in the stomach, with grave consequences: he never recovered from the blow, and died four months later, telling his confessor before his death that it was the abbot's blow that would kill him.

It is hardly surprising that the plaintiffs deprecated the abbot's meanness, cruelty, malevolence, and violence: they simply describe him as a *tyrant*.[40] The (allegedly) tormented family thus made their views about the abbot clear, but they do not reveal any *expectation* of such conduct from a monastic, as they would have done had they been full of anti-monastic malice against everyone with a tonsure.

What is the value of these cases as evidence? What can be deduced from them, since, after all, they are very few in relation to the number of religious houses? All in all, I do not see them as revealing widespread hatred against religious. Even when the plaintiff's life has been threatened, the victims (or their lawyers) are quite restrained. In a case in which the Abbot of Whitby was accused of buying stolen goods from pirates, the victim of the original robbery did not attack the abbot personally when he puts the case in writing.[41] The fact that the defendants were *abbots*, religious heads, seems not to have been relevant. Generally, the plaintiffs wanted justice because they felt they had been wronged, and it seems that there were no grievances – at least no voiced ones – against the religious orders as such. The documents do not tell the whole scene behind the short notes, but as we have seen, they do quote words characterizing the nature of the accused. It is worth noting that these cases do not condemn religious as such; rather, they condemn the actions of specific individuals. Although it may be argued that the criminal courts were hardly forums for voicing anti-monastic criticism, I firmly believe that because the juridical processes were based on *appeals* to the king, people had no reason to suppress their resentment. I am therefore convinced that people considered cases like these to be discrete incidents, and not evidence of the rottenness of the entire monastic system. Perhaps this is a key to understanding the reactions to late medieval religious and to religious life in general.

So far I have focused on the question of monastic lands and landowning. Land was by no means the only source of income for religious houses, but the emphasis here reflects early sixteenth-century concerns. The question of land loomed over monastics as if it had been their only form of income, and other important sources, such as industry, churches, and tithes, were often forgotten by their critics. The fact that religious houses ran industries such as saltworks and mines, fisheries and mills, and sold their produce, received much less attention than questions about landownership.

Religious houses were big employers. They provided work for artisans of all crafts: sculptors, masons, blacksmiths, tilers, glaziers, and so on, and not only for new building; craftsmen were constantly needed for rebuilding and repair, since the monastic rule obliged the abbots to keep their buildings in good condition.[42] When reformers commented on the significance of monastic employment, they typically argued that monastics employed people only because of their own idleness and reluctance to occupy themselves in manual work. It is no doubt true that the grandest abbots had servants and lived like the nobility (of course, they sometimes *were* of high birth), but for most, the conditions were much less splendid.[43] In 1538 Butley Priory, which is a good example of an average house, housed twelve Augustinian canons and two chaplains. In addition, the household included ten keepers of the land,[44] and eleven men in the kitchens, bakers and brewers included. There was also a barber (tonsures required frequent services of a barber), a butcher, five shepherds, a blacksmith, a gatekeeper, a gardener, five carpenters, twelve workers, one swineherd, six washerwomen, a schoolmaster, and seven children to be educated (they were supported by alms). For various other tasks twenty further people were employed. Butley thus offered a home or part of an income to at least 96 people, of whom several could be counted as charity cases. Against some allegations, this case shows that monks did not have personal servants.[45] The people who were employed at Butley worked for the day-to-day upkeep of the house, not directly for the personal luxury for the canons. Even though the household of Butley was ordinary and unexceptional, it offered sustenance for dozens of people.

Religious houses were a temptation to the gentry and nobility when it became clear that they would be dissolved. The houses had the reputation of being relatively well maintained and their lands were productive. Many gentry were keen to buy the lands from the Crown, or better receive a piece of land as a gift from the king.[46] This opportunism aroused the criticism that the very people who had attacked the wealth of the monks were now greedily snatching the monks' possessions, without a thought for the fact that hospitality and schools were being destroyed, that the rents of tenants on monastic lands were being raised, and that the peasants suffered when the monastic lands passed into lay hands.[47] Indeed, peasants were not necessarily pleased when their monastic landowners were replaced by laymen. The peasants were accustomed to coexistence with the monks, to 'peace and quytnes,' and were reluctant to accept new ways.[48] Warnings were soon heard that the new owners would raise rents and set new terms on tenancies. The peasants' complaints about the change of ownership speaks of approving of monastic landowners rather than of condemning them, and this argues strongly against anti-monasticism.

It was not only peasants who saw the non-monastic future as gloomy. Some envisioned a Lutheran future which resembled hell. Thomas More, for example, wrote that the heretical religion would lead to unruliness and complete misery in ungodliness, which would be followed by crime, idleness, beggary, and disobedience:

After that the clergye ys thus destroyed and caste owt / then shall Luthers gospell come in / then shall Tyndallys testament be takyn vp: Then shall false heresyes be preched: Then shall ye sacramentes be sett ate nought: Then shall fastyng & prayour be neglected: Then shall holy sayntes be blasphemed: Then shall almyghty god be dyspleased: Then shall he wythdrawe hys grace and lette all runne to ruyne: Then shall all vertue be hadd in derysyon: Then shall all vyce reygne and runne forth vnbrydeled: Then shall youth leue labour and all occupacyon: Then shall folk waxe ydle and fall to vnthryftynesse: Then shall horys and theuys / beggers and bawdys encreace: Then shall vnthryftys flok togyder and swarme abowte and eche bere hym bolde of other: Then shall all lauwe be laughed to scorne: Then shall the seruauntes set nought by theyre maysters / and vnruly people rebelle agaynst theyr rulers: Then wyll ryse vp ryflyng and robbery / murder and myscheyfe / & playn insurreccyon / wherof what wold be thende or when you shuld se yt / onely god knoweth.[49]

Much of the despair in More's portrayal is caused by the devastation of the church lands: poverty and misery quickly spread. In short, the world would be turned upside down. Chaos would reign. If More's vision sounds familiar, it is probably because of the vision of Cuthbert Tunstall discussed earlier. Tunstall feared that England would be invaded by the pope; the terminology and the visions of fear are similar.

More wrote some ten years before the final blow to the monasteries, and naturally could not have had any notion of the dissolutions. Therefore, his vision is curiously seconded by a devout evangelical in 1542, well *after* the dissolutions. Henry Brinklow, an ex-Franciscan and a married man in exile, regretted the loss. Even though he did not want the religious houses restored, he admitted that they had not been completely wicked. Many disadvantages of the dispersion of the monastic wealth were now apparent. The new owners of the monastic estates had raised the rents, and Brinklow wrote that it would have been better for religion in general and for the common weal if the land had remained in the hands of the religious. Under new masters the lot of the peasants had deteriorated; they had become all but serfs. The old masters had at least been bound to give alms and offer hospitality when times were hard. Brinklow commented that everyone knew that monks had helped thousands, and, furthermore, religious houses had always appointed a priest to take care of the parishes they received as benefices. After the monks left, no one hired a good priest or gave a halfpenny in alms.[50] Brinklow clearly felt frustrated about the outcome of the dissolution. His words sound an echo of anger about facing a world of lost and wasted opportunity. His anger sounds even more formidable if we remember that he was, after all, an evangelical reformer and a former friar who had given up religious life.

But Brinklow was not the only disillusioned reformer. An anonymous author's treatise demanded in 1544 that the income of the monastic lands be directed to schools and helping the poor.[51] Clearly, this was not the case. Bishop John Bale wrote that the monasteries were important monuments and strongly disapproved of his countrymen who would allow them to be destroyed because of some people's greed. Bale was Lutheran in his orientation, however, and refused to make any concessions regarding the justification of the suppressions. He saw the dissolutions as right, but regretted some of the consequences of the king's actions:

> I must confesse them most iustly suppressed, yet this would I haue wyshed (and I scarsely utter it wythout teares) that the profytable corne had not so unaduysedly and ungodly peryshed wyth the unprofytable chaffe, nor the wholsome herbes with the unwholsome wechs, I meane the worthy workes of men godly mynded, and lyuelye memoryalles of our nacyon, wyth those lawsy lubbers and popyshe bellygoddes.[52]

The fact that Bishop Bale deplored the wasting of the religious estates puts the theory of anti-monasticism in a very strange light. The dissolutions are regretted much more than the theory would admit. Religious houses are not remembered only as corrupt, vicious, and hypocritical: they are remembered with a sense of great loss and perhaps even with some guilt.[53] Especially after the long research tradition which has stressed the Protestant experience and general ill-will toward the popish Church, these representations are not what one would expect to find.

I do not think it far-fetched to suggest that for centuries the wealth of the Church had caused mild irritation, but it was nevertheless tolerated. It must be remembered that it was ordinary men and women who donated the monastic wealth. The evangelicals, however, had incorporated the issue of monastic wealth into their program against the papal Church, following the native example of the Lollards and especially the more recent one of the German Lutherans. Arthur B. Ferguson has suggested that there was a proper (Lutheran) campaign against the wealth of the clergy in England in the 1520s,[54] but this – especially the term 'campaign' – seems exaggerated considering the small number of active anti-clerical writers.

Monastic land was important: why else would the Crown be interested enough to go through the process of confiscation? The dissolutions were acquisitions of land much more than the implementation of new doctrines. Monasteries simply had too tempting a fortune and the tide had turned against them. Without their land and buildings, religious were unable to live in their communities and were forced to disperse, even though religious life in itself was not prohibited.

How was the wealth and poverty of religious life perceived? It is certain that the virtue of having nothing was a part of the religious pursuit and of the early modern way of thinking. It can be quite safely claimed that the English understood the nature of monastic ownership: namely, that individuals owned nothing. For everyone, land held essential symbolic value: more than anything else, land symbolized bread and life itself. Land and soil supported life in every possible way imaginable. Without land there was no bread, no meat, and probably no means of buying them, and no hunting and fishing rights. If land was so important, why did people donate portions of their own share of the land to the monasteries? The answer is simple: people wanted security in their life after death. True, donations of land were fewer in number in late medieval England, but the explanation for the relative scarcity of donations is not necessarily a weakened interest in monasticism or votive prayer, but in the fact that there already were so many established houses that the country was simply saturated. No new houses were needed. Second, the new orders, which required fewer donations when they settled in England, were fashionable. These two reasons give no ground for assuming that people had become anti-monastic. They had not ceased to support religious in their wills, nor did they stop recruiting. On the contrary, people still valued the donations that their ancestors had made and tried to take care of the monastic foundations of their predecessors. The land given to monks was the soil on which prayers for their own and their descendants' souls would be cultivated. More monastic land yielded more prayers, and more prayers further eased their way to heaven. Land was a sign that kept men alive in this life, and likewise their souls alive in the next. Monastic lands were important *because* they were monastic lands: in them were planted the seeds of nourishment not only for the earthly vessel, but for the numberless souls in the next life as well. The power of such a mentality should not be overlooked.

5
Crystal Palaces of the Voluntary Poor

Fishponds, fields, mills, and meadows constituted only part of the monastic wealth; religious houses were often grand complexes of buildings, adorned with enormous treasure: jewels, gold, and silver. The buildings included luxurious features: for example, much emphasis was put on hygiene, with efficient indoor latrines with running water. Many of the churches were magnificent, and there was competition between monastic and parish churches in building ever more beautiful churches and elaborate shrines and altars. Nothing was too good to honor God and his Son.

The precious stones and gold alone were enough to offend some who sought more austere forms of monastic worship. For centuries there had been a sentiment within the monastic orders that their treasures should be limited or that they should at least no longer be allowed to accumulate. For example, the Cistercian order was born out of the need to purge the monastic life of excess luxury and to search for purer forms in devotion and appearance. The monasteries they built were at first very simple (although often grand in size) and bare of decorations, but in the course of time, especially in the fifteenth century, the English Cistercians, too, surrendered to the riches they had received from their devoted supporters. Building beautiful churches and houses was a practice favored by abbots who wanted their names in history, and the austere structures of monastic churches and chapels tended to become enriched according to the decorative architectural fashions of the fifteenth and early sixteenth centuries. The large Cistercian houses in England, such as Fountains or Rievaulx, are examples of this development.

The religious of the early sixteenth century were capable of self-criticism, however, and knew that riches could become snares and slowly eat away at the very foundations of their vocation. Richard

Whitford, for example, wanted to set limits to the decoration and adornment of monastic churches and other buildings belonging to religious houses. He urged his readers to remember that the Devil himself was lurking near those who felt excessive pride in great houses and other riches. He accepted that God and Christ deserved the best, but pride in polished gold and silver in the name of Christ was dangerously close to pride in the name of Lucifer. Whitford hinted that those who felt earthly pride in their golden churches were like the children of Israel who had worshipped both God and their pagan idols after their flight from Egypt. The confusion would be too great, he knew, if men built for their own honor. Moreover, this certainly would not have been missed by the Devil. Everything was to be done to God's glory only, not to the honor of humankind.[1]

Despite these protests, the late medieval church was characterized by splendor. Even the smallest chapels would have several images and a lavishly decorated altar, and it was the parishioners themselves who acquired them for their local churches.[2] This was voluntary activity; collections for the parish churches were organized constantly. Guilds sponsored their own churches and saw to it that they were no less magnificent than the one on the next block. Even the poorest parishes kept their churches in as good a condition as possible. These activities were important for the individual, because it strengthened his or her role in the community. If we consider what an organic whole the people, the parish, and everyday worship was, the way people behaved, created, and recreated the role of religion in their lives becomes more easily understandable. In this context, it is no wonder that religious wanted to pay their houses similar attention. For monks, lavishing the religious houses with gold and silver or their altars with precious stones was not a mindless waste of money but a rational investment in the future and in pleasing God.

Riches became dangerous when things got out of hand and when there was a superabundance of everything. There was always the danger of an individual becoming interested in outward things, riches instead of God. With his critical eye, Richard Whitford not only found some churches and chapels too lavish, but also expressed his anxiety for some abbots, priors, and abbesses, who lived in too luxurious quarters. They had large, well-heated houses for themselves; he was afraid that a life lived in this kind of luxury would lead them to use the property as if it was their own. Whitford was resolute that this was not the type of poverty Christ had exhorted his followers to take upon themselves.[3] It is interesting that Whitford writes that it was dangerous for religious heads to regard common property as their own. This strengthens the notion we discussed earlier: collective riches did not compromise individual poverty and were thus acceptable and favored by God, whereas personal riches put one in grave danger and threatened the soul. Collective riches were permissible – provided they stayed collective.

There were many, however, who thought that such collective riches were unnecessary. Some commentators found the riches of the Church an embarrassment and argued, assuming that the total amount of riches was constant, that what the religious held was thus taken away from the genuinely poor (as opposed to voluntary poverty). People like St German urged religious houses to give up some of their possessions; they could begin by discontinuing the sale of indulgences and the collection of fortunes from pilgrims. Monks and nuns should focus on fasting, praying, wearing hair shirts, giving alms, and doing good deeds for themselves and others. This they should do with humbleness, 'cryenge continually to our lorde,'[4] and if they thus concentrated on the essence of the religious life, they would arrest the decay of monasticism. Erasmus, for example, thought that if gold had not been allowed to play such a major role in the religious life, monasteries would not have attracted fortune-hunters, people who professed precisely because they wanted to escape poverty. Such people were a serious threat, because once professed, they would be guided by their carnal lust and, in the end, obey no one.[5] Wealth and riches could be an extremely dangerous combination for human souls.

Some evangelicals attested that individual monks were greatly affected by riches and took pride in the grand settings of their lives. This could make them forget their place in the world and think too much of themselves, even to the extent that they perceived themselves to be better than noblemen. So proud were they that they hardly let 'a Lorde of bloode with them to sytte.'[6] The most militant evangelicals stressed that there were no worldly princes who could compete with the Church in the glory of castles, because the most pleasant 'dwellinge houses' were in the hands of the clergy.[7] Their houses were beautiful and clean, their walls strong, and their land the most fertile.[8] The evangelical message is clear: if the clergy built a paradise in this world, this indicated that they were no longer concerned with the afterlife. Some of these critical comments are not merely contemptuous, but envious of religious. The clergy were criticized for caring *too* much for their buildings, keeping them clean and in good condition, which for religious, however, was an obligation set down in the monastic rule. Accusations of building a paradise on earth have a ring similar to the ancient notions of the monasteries being a sweet smell to God and an earthly Jerusalem. The critics did not accept that all the riches were intended first for God and only second to remind people of the glory of the life hereafter. From the orthodox point of view, everything was there for a reason, because the sensual nature of worship required these symbols of the divine. Their grandeur reminded people that God's glory always surpassed earthly luxury and therefore made tangible the future he offered. Luxury in itself was not an end, but simply a means to reach God and to speak to him.

The evangelicals found the aspiration to demonstrate adoration of God with glittering objects despicable and a sign of abandoning God and his truth. Here, the clash between the reformers and the conservatives seems profound, since both parties set their beliefs on foundations with very long traditions. Simon Fish, for example, accused nuns of embroidering expensive altar cloths and curtains to decorate monasteries and chapels, although such textiles were 'all nothynge but pryde / pompe, and concupyscence of the eyes.'[9] For those accused, such accusations must have seemed misconceived. From the nuns' point of view sewing and embroidering were their most important occupations and an essential way for them to show their love for God. It was their labor and duty. Fish talks about the folly of decoration, but the majority of the English probably did not understand his point, since not only religious, but also lay men and women were voluntarily occupied in such activities.

In the early modern mentality, there were not many signs of luxury stronger than crystal and glass. They were expensive and extremely pleasing to the eye. Only the rich and the powerful could afford large quantities of glass, and it is no wonder that in poems and ballads, for example, windows were a sign of wealth. In a pleasantly built environment, crystal played a part in creating a peaceful, harmonious, beautiful, and luxurious atmosphere.[10] Being devoted to God who loved beauty, monasteries naturally were famous for their glass. In popular stories monks and nuns lived amidst lapis, coral, emeralds, onyx, amethysts, topazes – and crystal. The cloister walls would be adorned with glass windows, and the brightness of crystal ensured that monks lived in the light.[11] Glass and crystal were important symbols of light, a commodity not self-evident in an age of thick walls and small windows. It is hardly surprising that, for example, the monks of *Peres the Ploughmans Crede* surrounded themselves with gold, alabaster, marble, gemstones, and glass, to such extent that their chapterhouse resembled a church or the Houses of Parliament.[12]

As glass was an important symbol of wealth and a way to serve God as well, it is not surprising that glass also appears in texts that criticize religious houses for their excess. Occasionally, the early sixteenth-century evangelicals saw precious stones, silver, and gold as symbols of the evil of the clergy and abbots; their heaviness lay upon them as a burden which the misguided themselves could not see. The crosier, it could be said, was so heavy that the hands of its bearer became crippled and could no longer perform charitable deeds.[13] These allegorical words strongly suggest that the cornerstones of religious life were lost beneath the opaque wrapping of wealth and love of riches.[14]

Regardless of the abundance of luxurious items in some religious buildings, luxury was not always material and shiny. Even heating was perceived as a luxury. Critics blamed the friars for indulging themselves in warmth and comfort despite their vow of poverty: when it was cold, they would wear socks and have a fire, and when they traveled, they went to the houses of the rich so that they could enjoy comfortable surroundings. (As an aside, it should be noted that monks were more often blamed for imposing on poor peasants by asking for lodging and food.) Poverty and the refusal to handle money meant nothing in the eyes of their critics if friars drank from golden cups, washed their feet in herbal water, and touched coins with their hearts.[15]

Next, a few words need to be said about the views of those who did not find all monastic riches dangerous. We have already noted that the primary motive for adorning churches was service of God and in that light it is hardly surprising that voices were heard in support of this practice. The wealth and prosperity of the monasteries were not always considered signs of diabolical living within their walls. On the contrary, it was common for writers, especially describing conditions in foreign countries, to praise the marvelous churches and abbeys and their price-less shrines.[16] This is clearly what ordinary devout people appreciated. Likewise, the founders of the richest monasteries were admired for their contribution to advancing the service of God. A rich benefactor or a founder gave religious the ability to practice their religion without worrying about survival. In a religious culture in which liturgical objects and vestments were necessities, a generous founder was praised for donating crosses and precious stones, vestments and vessels, books and land.[17]

Bradshaw saw no item as useless, no treasure too expensive or lavish in God's eyes. Donations served both God and the security of the religious, both financially and physically, by keeping enemies outside the fortified walls. Henry Bradshaw admired St Werburga's father, who according to legend had founded Peterborough Abbey and Stone Priory as a penance for his sins. Founding a *rich* abbey and endowing it with valuable rents, liberties, and other possessions, thus offering a suitable place for religious to serve God day and night, brought the king nothing but merit. Confirmation that founding these houses had served this specific founder's interests well was later provided by a series of miracles that took place in them;[18] no house could wish for more, and for anyone sensible, these miracles proved that the foundation was justified and that the founder had done excellent service to God. Homage was what God expected.

As we can see, wealth was not always regarded as harmful. Many argued that by giving God riches one could legitimately acquire merit for one's own or someone else's soul. Prosperous times and prosperous monasteries are thus inescapably associated, for wealthy monasteries benefit both science and virtue:

> Science encreased true loue and amite
> Vertue was exalted in all this region
>
> Monasteries were edified of his benignite
> Endowed with riches / and riall possession.[19]

Sir Thomas More used the same arguments against his enemies the Lutherans, and further proved that not everybody thought grandiosity was evil: he too thought it justifiable to build beautifully adorned churches in God's honor and glory.[20] Furthermore, it was not fitting to serve God with metals less valuable than those used in secular contexts. It would have been the sin of Cain to keep the best portion to oneself and present God only with things that were of lesser value. Practical considerations also had to be kept in mind: More remarked that wooden chalices, for example, were simply impractical for storing the blood of Christ, because the holy fluid would be absorbed into the material, which was no less than sacrilege.[21]

Donations, lands, and chattels given to religious houses were in the end given to Christ and in his glory: they were his patrimony.[22] Therefore, interfering with them was against God's will and such iconoclastic tendencies were strictly condemned. More wrote that the iconoclastic evangelicals were not merely criminal, but much more witless than one of his fools, Clyffe, who broke an image of the Virgin Mary on London Bridge without feeling remorse. Clyffe was astonished at the fact that the image had not been repaired afterwards; in Clyffe's logic – and More was a man who understood folly – the guilty were those who did nothing to repair the damage.[23] More wrote that his enemies were not nearly as sensible as Clyffe, because they could not see the truth, even though it was plain and simple.

The accusation that monastics surrounded themselves with wealth suggested that many people envied contemplative religious life; the reformers assured anyone who felt the pangs of envy, however, that the religious life was not easy. Thomas More was certain that most people would return running to the world after trying to live a monastic life. Many complained and cursed their life when their wives were angry with them, but if given the opportunity, would not change it for the more difficult life that a monastery would offer.[24] Life in a monastery was not one of luxury, nor was the light of crystal or the brilliance of precious stones sufficient reason to enter such a demanding life. Abstinence meant that whatever luxury there was, it could not be truly enjoyed by the monks, who had to concentrate on their pious exercises. Anyone who envied religious for their riches, good food, and lack of physical labor, would soon see that life in a monastery was not worth envying, and repent the loss of their liberty.[25] More suggested that for most people luxurious items could never replace the loss of liberty, that is having to give up one's will and replacing it with complete obedience.

What the religious houses in fact possessed seems far from the insinuations of the wildest claims. During the dissolutions, it was not always gold and silver and huge quantities of lead that were carried away. From some houses it was difficult to gather many items worth selling. For example, the London Charterhouse yielded stone, wood, and panels. The king received herbs, trees, carp, timber, stone, lead, and glass from there, and Thomas Cromwell three baskets of herbs. Others were given herbs, hay, carps, cloths, and a *mappa mundi*.[26] This kind of treasure did not come close to carrying away cartloads of gold and precious stones.

The redistribution of the wealth of the religious houses did not command wide approval, for many felt that religious had a right to collegial ownership of things which they acquired by working or as donations. Similarly, respect for ancestors demanded that the gifts they had given should be honored. Everyone, including many religious, agreed that excess wealth was not good for the religious soul. The evangelical claim against the possessions of the Church was nothing new; it was rather the continuation of a tradition which reformers both inside and outside of the religious orders had long supported. Even those who approved of golden churches accepted that too wealthy a clergy was not necessarily beneficial.

No consensus was found. The issue of monastic wealth was extremely difficult: even if the monasteries had been too rich, the dispersal of their wealth caused widespread distress. Although the churches and buildings were lavishly decorated, the people would not, in the end, have it any other way – it was, after all, the way things had always been done. Everyone knew that most monastics had reasonably comfortable conditions, and by endowing them suggested that they did not altogether disapprove. The success of the mendicant orders does indicate, however, that poverty was regarded as holy. In conventional thinking, this problem was resolved by collegiate ownership. There were at least in a technical sense no rich individuals in monasteries, and despite the danger of feeling pride in their well-appointed houses, monastics' souls were prepared to enter heaven.

Part III
Castitas

6
Mortifying Monastic Flesh

Religious persones ben more streytly bounde to gyue hede /
and to haue them selfe with more deligency /
in awayte for the custody and kepyng of chastite /
and to be more carefull and dredfull of the losynge
and breakyng therof: than any other persone.

Richard Whitford: *The Pype*, f. 206v

For a good Christian chastity was essential. Chastity – or cleanliness – was an interplay of mind and body, since the body could not stay chaste without reason, and to keep the mind clean bodily control was needed. Part III discusses aspects relating to keeping one's mind and body clean. This chapter discusses the role of the body in early modern religious culture, and in the following three chapters we will look at outward appearance, feeding the body, and finally our interest will be aspects of sexuality.

A human being, however holy and humble, was always a dual entity,[1] a mortal habitat of flesh, slime, and sperm, a detestable dwelling place for worms as well as an eternal soul which might one day join the angels in their faultless beauty in celestial joy. To distinguish between the soul and the body, it was imperative to recollect that man had been conceived in filth and shame. The body was unclean, it was 'fouler than ony donge or slyme,'[2] and it was 'a stynkyng sperme or seed, a sacke of dunge, & the meat of wormes,'[3] or an 'vnclene carcas.'[4] To overcome the shortcomings of the body, one had reason. Animals as well as humans had senses, the five wits, to coordinate one's needs, but it was reason that separated man from beast.[5] Human beings were considered exceptional creatures in the sublunary world because reason enabled people to control their actions and pursue something higher than reason: love of God. It was relished with joy that of all his creations God had especially chosen humans as recipients of reason so that they could enjoy all his gifts. A circle followed: by seeking God and his love humans could, with God's help, employ their reason to control their senses and their bodies, become as little beast-like as possible, and reach their ultimate goal: sanctity and even God.

Late medieval and early modern people seem to have been almost obsessed with the danger of bestiality. Animals were inferior creatures: they had nutritive and sensitive souls, but not the intellectual soul, which only humans possessed. This gave humans a responsibility not to follow only their bodily needs. This responsibility meant for many great sacrifices, which were nonetheless supposed to be made willingly. Because the flesh was the closest thing in man to the animals, it required special attention. It was commonly acknowledged that the flesh was dangerous, false, and treacherous, and that it could deceive even the most pious.[6] The poststructuralist notion that culture is experienced with the body applies here.[7] Uncleanliness and beastliness went hand in hand because all carnal knowledge was transmitted by the senses. All the senses had their dark sides: the sense of smell, for example, nurtured superfluity, and touch was the enemy of chastity.[8] The sense of touch could be more dangerous than sight, because this sense was powerful enough to divert the mind, upset it, and kindle unlawful affections. This is why monastics were not allowed to touch each other. The sensation of touch could bring the mind 'vnto a soden furie or madnes,'[9] and in such a madness, even a wise man could lose his reason, become beastly, and be tempted into sinful behavior. Because of the dangers they presented, the senses had to be either extinguished or preferably opened to receive only God's love. To achieve this, control was necessary.[10] Religious houses were intended for the practices of control. Proper cloistering, shutting oneself out of the world of sin and sorrows, made the monk a spiritual eunuch who avoided the easiest traps of the world in order to struggle with more tremendous foes. Chastising the body was necessary in order to purge the senses and subdue the flesh,[11] to purify the mortal body from earthly mud and to prepare it for what was to come in the after-life. Treating the body harshly served a wonderful purpose: a mortified body would be free, and freedom, of course, meant liberty from sin. The body existed only as a tool for attaining liberty.[12]

If this liberation was not reached, the outcome was the soul's eternal captivity. If the treacherous senses were not kept under control, sensuality was let loose. Sensuality was compared to a well from which sprang 'fayre water and swete,' but which had a dark bottom from which displeasure surfaced causing people to avoid fasting, vigils, cold or heat, and all perils.[13] Sensuality was also the 'very bondage & moste caytyfe thral-dome.'[14] In the religious life, this exhausting battle was the heart of the matter: the dark bottom of the well had to be avoided, but this could be done only by learning to accept things that were not pleasant to the flesh. It was not enough to repent every now and then; one's whole life had to be focused on penitence. Abstinence had to be ceaseless to benefit the soul at all, since, for example, excessive eating and drinking followed by strict abstinence only inflamed the senses. The situation was not, how-ever, hopeless. Sensuality could be overcome, but only by often drastic measures; the flesh could be tamed only if sensuality was 'made bonde captyue for euer.' Only then would dullness, sorrow, doubt, despair, and troubles be replaced by joy, trust, quietness, and rest.[15] The senses could be made prisoners only by spiritual growth for which mortification of the body was essential. Caroline Walker Bynum notes that the body is 'not so much an hindrance to the soul's ascent as the opportunity for it. The body is the instrument.' Bodiliness could provide 'access to the sacred.'[16] Without a sinful body it would have been impossible to reach God, and even in its cruelty mortification was wonderful, because the reward was 'pleasure and gladnes.'[17] Nothing could be achieved without suffering, but salvation was a great motive: if one gave God hours and hours of suffering, he would give a sweet reward. If one extinguished one's senses and observed God's word diligently, one would bear the true cross, the true yoke of Christ.[18]

Suffering permeated the whole of medieval religious culture.[19] Pain and torment were as important physically as the notions of sin and virtue were mentally, and achieving virtue and thereby a good life was impos-sible without suffering. In order to undo original sin, every believer had personally to take part in the suffering of Christ, his Passion. Taking the cross of Christ upon one's shoulders was a noble and wise act, and an essential element of practical religion. Because the cross would help people forget their worldly occupations, it would in return grant them freedom and liberate their spirit.[20] The ideal surrender to the Passion would be so total that a person could consider entering a monastery.[21] The model of an ideal life had for centuries been the monastic way of life. It was generally accepted that monasticism was the form of life that could best embody an answer to the call of Christ. Christ's passion had been bodily torment, and only pain could make his followers truly aware of his suffering:[22] 'The state and perfection of Christes religion / dothe requyre violence and stryfe.'[23] Because Christ had suffered so tremendously, and because his Passion had to be imitated in the hope of salvation, no measure was too drastic as long it did not end in suicide. Consequently, a penitent was willing to suffer all kinds of torment. In this respect the suffering body of Christ was as interesting to the sixteenth-century devout as to the medieval mystics. Therefore, Christ's passion was untiringly recreated, often in extremely realistic and grotesque terms,[24] as in Bishop Longland's sermon of 1536, which describes Christ's suffering in the hands of his tormentors. Bound to a pillar, he was scourged and crowned with a crown of thorns. His hands and feet were pierced with nails. He was stretched on the cross, which made his veins and joints crack so that his bones could have been counted by the sound they made. Another torment began when his cross was raised and hammered into position.[25] Cuthbert Tunstall stressed that death on a cross was far more painful than death by hanging, drowning, or from a blow to the head.[26]

The ideal body seems to have been a grotesque body.[27] Like Longland's description of Christ's cracking joints, Thomas More wrote about the agony of Christ's heart as the apprehension of his violent end made his body sweat blood. More's vivid images of the *gourging*, nailing, stretching of the limbs, and the blood running down Christ's face created a strong contrast to the image of the beatifying religious fervor in which he committed his soul to his Father. To everyone who truly believed in Christ, all this would prove that no prayer uttered in earthly pleasure and comfort was as effective as those made during great suffering.[28] Later in his text, More added to Christ's suffering mockery, scorn, and shame, and stressed that suffering all these was worth while since it led to everlasting glory.[29] Christ's fate seemed extraordinary and his suffering overpowering because his body was perfect and his soul stainless. No mere mortal's pain could correspond to Christ's, because his suffering had decided the fate of mankind; in this sense he died millions of deaths. Elizabeth Petroff posits that Christ's pain in a way transforms human pain: 'The surrender to pain cannot become transcendent unless the visionary sees that Christ's pain is of a totally different order than normal human pain, for it is not so much physical pain, grief, or loss, as it is love and salvation.'[30]

The broken body was a spectacle, a horrible sight to be chastened by.[31] Christine de Pizan repeats Boccaccio's story of Argia, a good wife who finds her husband's body on a battlefield and kisses it, not letting 'the infection of his body nor the filth covering his face . . . stop her.'[32] This is not very far from the treatment Christ received in contemplative imaginations.

The body was not supposed to enjoy the violence it was subjected to: the crucial matter was the pain itself. Sixteenth-century culture was permeated by the idea of healing and educating by means of violence, but asceticism of course was a much older ideal.[33] Apes and dogs were taught to dance, against their nature, by beating, and children were brought up with the rod because they were thought to learn best with its help.[34] Likewise, it was against the nature of men to resist sin, but suffering would teach them the right way. It follows that the medieval religious understanding of pain was holistic: it was educative and frightening, comforting and depriving, all at the same time. And pain was necessary. Similarly holistic was the notion of the body. The medical and theological theories of the body and soul did not differ drastically: both were a combination of the body and the soul, and both were affected by the interplay of the mind and flesh. The body's health reflected the soul's health, and vice versa.[35] Therefore, what was done to the body had an effect on the soul.

It is now time to discuss the ways in which the body could be used in order to improve the mind. The most powerful was perhaps mortification of the flesh by punishing it with different practices and devices, but at least equally important were the less dramatic tools of contemplation and labor. First, a few words on the practices of punishment of the flesh. I will take only one example – hair shirts – which we know were quite widely used even in the early sixteenth century.

Hair shirts were used to mortify the flesh. The coarse hair ultimately broke down the skin and caused great suffering. The wounds could become infected; if one aspired to more pain, one could lie on a hard board for a long period wearing the hair shirt. When the flesh began to rot, the ultimate point in bodily mortification was soon reached. By no means were these methods unique to religious, but the Carthusians were renowned for wearing them under their habits at all times. Thomas More, a long-time friend of the Carthusians, respected the practice, but found dangers in using hair too rigorously. In order to limit suffering, he wrote that not all religious should wear hair continuously, because it was so consuming.[36] Evidently he thought that there were other ways of expressing holiness, and that it was unnecessary for all religious to follow an almost inhumane example. More did not think bodily chastisement was always hypocritical as did the evengelicals whom he so strongly disliked. He actually reproached them for thinking that bodily chastisement was always hypocritical:

> You think all obedience of the body offered to God is hypocrisy. But it was not hypocrisy to Mary, who washed Christ's feet with her tears and dried them with the hair of her head. Coarse clothing is hypocrisy to you, but it was not hypocrisy to John the Baptist, who dressed

in the skin of camels. Abstinence from food is hypocrisy to you, but it was not hypocrisy to John, who ate only locusts. And it was not hypocrisy even to Paul, who wanted to be able to fast all day.[37]

More could hardly have had more influential authorities for his cause than St Mary, St John the Baptist, or St Paul, who all, according to tradition, had practiced at least one form of mortification. For More, the heretics themselves were hypocritical, because they talked about adoring God 'ardently in spirit,' but did not feel their flesh becoming hot when they worshipped him.[38] To honor God properly, something physical had to take place, since people were a unity of body and soul. Both the mind and the flesh had to move. If the heretics were left cold it only proved to More that they were not able to communicate with God.

It thus becomes evident that the importance of the body in early sixteenth-century culture should not be underestimated. It is apparent that focusing worship on the soul was still quite new in early modern faith. It seemed unfamiliar and odd to many that Lutherans especially had become estranged from the flesh and their bodies in their religious experience. In this sense, their religion had a different focus. When evangelicals talked about the suffering of the body, they talked allegorically, whereas supporters of the old order still meant living flesh and blood, in very concrete terms. Yet the evangelicals and the Catholics used the same language about the body and employed the same metaphors. This is also noted by Natalie Davis in her study of the Reformation in Lyons. The language of the two rival groups was still shared during the 1530s.[39]

The flesh was in its most fallible state when the body was idle. Idleness as a concept generally indicated destruction and decay of all religion. The idleness of the body or the mind was equally perilous, since a lazy mind would create sinful and lustful thoughts, which then easily led to the destruction of even the most chastised bodies. The appetite of the flesh dangerously burned and stirred those who were lazy.[40] Perhaps it was because of the dangerousness of the idle body that spiritual exercises, contemplation, were integrated into some activity. This is well illustrated in an exhortation to nuns to take the role of the Virgin Mary and nurse a doll or a statue of Christ as a living child Jesus: 'ye may ioyfully clyppe ye chylde, kysse that chylde, ebbrace that chylde to your clene brestes. O now is this a mery medytacyon for maydens.'[41] This 'merry' exercise, very popular in the late medieval nunneries and convents, was not a substitute for real family life for women who would remain perpetual virgins – though in practice it may have been that, too

– nor childish play for deprived adults, but rather a way of feeling holy emotions and searching for mystical experiences. Several important religious benefits would follow. First, by nursing the image of Christ, a nun could, literally, enter into Christ's life and his world. Second, she could express love toward Christ himself. Third, she could participate in and identify herself with the life of the Virgin Mary, the most perfect woman. Sharing her emotions and participating in her unique experience of immaculate conception and giving birth to God was no doubt a noble and aspired aim. Fourth, through this exercise, the nun could feel the perfect mystical experience of *caritas*, God's love. As we can see, meditation was also sensory and intellectual. The results mattered; the methods themselves were of lesser importance.[42]

When contemplation could be practical, work could be contemplation. Both were necessary, since idleness and laziness were considered utterly horrible sins. The idea that laziness was evil was a part of the mentality of early sixteenth-century people; it was no creation of 'the Reformation' – sloth was one of the deadly sins. For the sixteenth-century religious, the sin of sloth meant failure in fulfilling one's religious duties. No doubt people understood the connotations when the evangelicals sometimes perceived monastics as living in 'slouthfull ydlenes,'[43] or as a 'swarme of ydele religious.'[44] Quite simply:

> No frutefull work they vse,
> All honest laboure they refuse,
> Geuen wholy to sluggyshenes.[45]

Anyone who wasted time doing nothing was regarded as welcoming all vices and sins into his or her life.[46] Idleness was understood to be the mother of all sins, because it prepared the way for all other vices:

> For ydlenesse the grete moder of synne
> Euery vyce is redy to lette ynne.[47]

The poet Stephen Hawes's eloquent verses on the deadly sins convey a sense of the fear people were expected to feel when they faced even the thought of these sins. It was considered extremely easy to be trapped by pride, envy, sloth, gluttony, or lechery.[48] Hawes reminds his readers that God had given free will to his children to be used for the cultivation of their virtues. Free will could be used in order to avoid the worst sins, but, if misused and treated irresponsibly, it could lead people to the worst possible end, to the soul's destruction.

What the evangelicals appear to have wanted to see was monastics working hard, not lazing like 'brute beastes and swyne,' who refused any honest labor.[49] They did not regard monastic reading or singing as worthwhile work.[50] Thomas Starkey, for example, lamented so many religious burdening the earth and nourishing idleness with their riches. Their idleness generated nothing but vice and vanity.[51] The avoidance of monastic idleness was one of the reasons why some evangelicals found it sensible to turn monasteries into schools and hospitals, which would be useful and profitable for all citizens. As always, there were others who offered counter-arguments, and in their turn, accused the reformers of laziness. Apparently, during a visit to Germany, William Barlow had found the German reformers themselves guilty of the sin of laziness. He speculated that Lutherans and other heretics intended to live idly on the plunder of the churches and monasteries, and on the extortion of innocents.[52]

Labor, pain, and penance were intertwined. Idleness was detestable and the mind had to be continuously occupied with something useful or holy. Richard Whitford sums up the idea of labor in his *The crossrowe or ABC*, in which he presents his *Alphabetum religiosorum*:

By labour, payne, penaunce, & by many tribulacions haue all faythfull persons passed this lyfe. And by them onely / muste entre the kyngedome of heuen: and contrary ydlenes / ease and pleasure in honour and delycacy: dothe teache a man moche myscheffe and euyll, and bryngeth him at the last vnto the pyt of perdycyon.[53]

The labor which was demanded of the inhabitants of religious houses was found in their handicrafts, which were often impressive. Though popular literature scorned monastic idleness, there were occasions which proved that some held different views. George Gyffard, one of the visitors to the monasteries in the mid-1530s, was quite taken by some of the houses. Gyffard's duty was not necessarily to look for excellence, but that is what he found in a small monastery in Leicestershire:

VIII religious persons … lyvyng religiously, havyng such qualities of vertu as we have nott ffownd in no place; for ther ys nott oon religious person thear butt that the can and dothe use eyther inbrotheryng, wrytyng bookes, with verey ffayre haund, makyng ther own garnements, karvyng, payntyng or graffyng.[54]

Like the Leicestershire monks, good Christians never allowed them-selves a moment without something useful to do. Monastic idleness could be prevented by occupying the mind with pious thoughts and prayers, or by reading authorized works.[55] The Chester monk Henry Bradshaw explained that one of the principal reasons for his translating and writing the massive poem on St Werburga had been to avoid the dangers caused by idleness. His life as a monk was nearly filled with the daily routine of his house, but he also wanted to have something useful to do during the long, dark hours of the night when he would or could not sleep, and also during those regulated periods of the day when everyone had to be occupied with some useful activity.[56] Bradshaw saw it as God's will that all his creations should fulfill their duties and carry out their daily share of labor. Birds kept themselves busy, and people should imitate them. The health of the soul demanded useful labor; it was difficult to reclaim time wasted.[57]

There was a wide range of suitable occupations for religious. Most often monks were involved in activities such as writing, illuminating, painting, or even making musical instruments. The nuns would sew, embroider, weave, and teach children.[58] What was actually done was not as important as doing *something*. Ideally, being occupied day and night was what mattered, but spiritual activities were held in higher regard than bodily ones. Reading was especially important. If a monk became weary of it, he was instructed to 'refresshe' his 'soule by prayer, or meditacion, or in some other vertuous and charitable werkes.' Staying active was rewarded: 'so shall thy houre be shorte, and thy labour light.'[59] The long monastic days and nights would seem much shorter for someone who kept busy and avoided idleness. Whitford thought that nuns should have a similar attitude to spiritual activities as men, at least if they were learned enough. He stressed, however, that nuns should not stay in the company of men unless it was absolutely neces-sary. As an example he mentioned the occasions when nuns needed teaching or spiritual guidance.[60] Since Whitford was prepared to allow nuns to spend time with men if it was for the purpose of spiritual teaching, he clearly felt that nuns' spiritual development was as import-ant as monks'.

The evangelicals strongly criticized the non-recognition of manual labor as essential in monastic life.[61] Their feelings clearly supported equal hardship for everyone; they refused to tolerate monastics who, in their opinion, did nothing to support themselves; they claimed that the early monastics had been used to hard work, and that they, unlike their contemporary brothers and sisters, had been able to support themselves.[62] The problems related, once again, to the accumulation of property, which allowed religious to concentrate exclusively on meditation.

The question of laziness was directly linked to the question of whose religion was more worthy in God's eyes. Many late medieval religious movements[63] argued that the yoke of Christ could just as successfully be carried in the world and amidst family and friends as in a secluded religious house. Instead of preaching the necessity of rejecting the world, they saw sheltering more as a sign of cowardice in life and in Christ's eyes than as a sign of complete devotion. On many occasions it was the peasant who was then elevated as the virtuous opposite of the monk. It could be stated, as Thomas Starkey did, that a peasant could be as devout as a monk:

> This man I say is a perfyte religious man, though he be nether couered with sainct Benets cote, nor yet with saynt Francys: ye & though he be at ploughe and the carte, and labourynge the grounde, yet may he be as perfite in religion as the moste perfyte monk lyvynge in his cloyster.[64]

Being unafraid of bad weather and storms is a virtue of the plowman and the sailor. Clearly, it is nobler to suffer the pains of the world than flee from them.[65] Anything that God wanted, a person could aspire to, and achieve, without becoming a monk or a nun. In the eyes of God, ordinary life is quite as acceptable as the religious life.

For some, having monastics doing manual labor would result in 'a meryer world than now is,' but the conservatives, too, had some ammunition left. Examples were again found in 'greuous complayntes & murmuracions' from Germany, where artisans were furious because their income had seriously decreased after religious had taken up handicrafts. In the Low Countries, a similar pattern was detected: to the dismay of the commoners, religious had taken up spinning, weaving, and making linen. There was now more competition and, as a result, less work to go round.[66] It is interesting that the evangelicals took idleness as such a central target of their critique. They could, of course, have concentrated on what they saw as a hypocritical practice, namely mortification of the flesh. Work was, as the Weberian tradition maintains, an issue close to the reformers' hearts, and they seem to have considered that monasteries produced useless things. Laziness became an issue more than round-the-clock vigilance and mortification of the flesh in the monastic world; monks were said to be idle regardless of the fact that they were under an obligation never to allow themselves to be without some kind of occupation.

Doing nothing and being lazy seem to have been issues that divided people. The evangelicals did not find it necessary to concentrate their whole life on mortification or penance. For monastics, contemplation was an activity and legitimate business, and mortification work that benefited the soul. These two views, of the reformers and the Catholics, did not have much common ground on which to meet.

7
Wearing the Monk's Hood

Even though it did not draw envious glances for its grandeur, the religious habit became yet another, although perhaps not the most eloquently discussed, issue in the Reformation debate. Clothing and outward appearance are effective means of segregation, separation, or distinguishing oneself from others, especially from those who are lower in status. Naturally, the nobility and royalty had always distinguished themselves by grandeur and luxurious materials, but as the middle classes grew stronger and wealthier, they were able to adopt the fashions of the nobility. Professional groups also differentiated themselves by distinctive clothing or other markers. The clergy wore cowls and tonsures; nuns veils; craftsmen clothes that identified them with a certain social group; masters, journeymen, and apprentices wore clothes that confirmed their status within their groups. Ideally, one glance would reveal who and what someone was. This brought security and a certain amount of reliability to society. Just as crossing the boundaries of class (for example, a peasant wearing silk) could even be criminalized, cross-dressing was dangerous and sinful because it broke what was understood to be the natural order of things.[1]

Clothes were therefore by no means a trivial issue, and this was recognized, for example, by prosperous merchants and others of similar standing. They invested in outward glory so enthusiastically that people with more moderate minds warned about such vanity and foolishness: this indulgence would definitely lead to damnation.

The author of *Ye olde god & the newe* observed monks carefully, and understood the allegorical meaning of the religious habit:

Monkes.
The monke goeth well nere smothe, hauing all his heare in a maner quite shauen of / and couered also with a great cowll that nothyng

84

may be sene / his garment is sydedowne to ye fote, his hose eyther beynge grey or whyte to touche his knew: when he passeth be the alter, or by his superyour, then he lewteth & maketh lowly reuerence / he casteth his hoode farre at his backe, & so afterwardes with a tremblynge heed he goeth a part in to his owne place.[2]

For a religious, the habit served many purposes. First, it was a sign of piety, a marker of the change from worldly excesses to the simplicity of the religious life. One of the greatest signs of devotion was a rich person's commitment to swap luxurious clothes for the coarse garment of the religious. St Werburga, for example, gave away her diadem and royal purple, and wore a simple black garment instead of the silk and cloth of gold of her youth.[3]

The second important function of the habit was that it assured that pride in clothes led no one into temptation. Allowing religious to wear anything they liked could have initiated competition in which each tried to look more handsome than the rest. This would have been worldly and therefore unacceptable. If everyone dressed in the same way, there could be no competition: for that reason no differences in shape, length, width, color, price, value, or quality should be permitted.[4]

Third, the religious habit was self-evidently important in covering the sinful body, since nakedness and pleasure at the sight of a naked body was bestial. Shame was a necessary feeling for the descendants of Adam and Eve, who after the fall had understood the need to be ashamed of being naked in front of each other.[5] Culturally, nakedness was perceived as an abominable sight. More than layman's clothes, the religious habit was thought to help to veil the mortal and sinful body. Faces and hands could also be covered to protect against carnal lust, both against the monk's own lust and that of others for him. Nuns covered their heads with the veil, which imitated the bride's practice at her marriage. The veil was not only a sign of chastity (and obedience), but also symbolized the nun's spiritual marriage to Christ.[6]

The religious habit was also used against the dangers of carnality within the monastery walls. Religious fully accepted that the dangers of the flesh were everywhere; the threat did not come only from without the gates. A clean body, for example, could lead to delight in cleanliness. Therefore, it was not suitable for St Augustine's disciples to wash themselves, or their clothes, too often. It was clear that 'fayre and clene clothes shulde defoule and make the soule vnclene & fylthy.' The abbot's duty was to see that habits and monks were washed and bathed, but not too frequently;[7] after all, filth could be healthier for the soul than cleanliness for the body. Attention to the body was understood to be mortally dangerous. The body was best left alone, under the fabric of the habit. If we consider Mary Douglas's influential theory, this is one of the rare occasions when in western culture dirt overrides cleanliness and reverses the norm.[8] The body is such a taboo that even washing it is an obstacle.

All clergy, both regular and secular, were recognized by their distinctive clothing; the tonsure, the cowl, and the nun's veil were signs of the position of religious. All of these were intended as symbols and special signs of holiness, purity, and humility in God's eyes. But these signs appeared suspect to the evangelicals, who questioned the justification and even the very Christian nature of these special garments. Reformers argued that religious should have been content to wear modest lay clothes rather than making an effort to stand out in the eyes of God. Critics found distinctive clothing sinful, since they could not find in the Bible any reference to confirm the necessity of a special religious habit. The defenders of the habit, on their part, referred to the Old Testament, which often mentions priests wearing distinctive clothing.[9] Furthermore, they argued that the reformers were guilty of wearing questionable clothes as well. They would wear bright colors or fashionable cuts: 'they go I say dysguysed strangely from that theye were before, in gay iagged cotes, very syghtely forsothe, but yet not very semely for such folk as they were and shulde be.' It was far better for religious to wear the habit of their predecessors than to wear 'the ruffyan inuencyon of many gospellers in Germany.'[10] There was no way a discussion such as this could have ended in consensus. Arguments were also raised by the fact that different orders had different habits. This could be interpreted as an indication that the Church was divided into sects, which was not of God. If the orders are interpreted, as John Frith did, as sects, the habit could be seen as the Devil's symbol, marking its wearers supporters of Antichrist and suggesting that perhaps the sects were his creation, not God's. Frith suggested that further destruction was created by persuading people to put their faith in this deception.[11]

In defense of religious, many argued that all Christians dressed differently. Once again, we come to the notion that hierarchy was the order that ruled the early modern mind. Hierarchy showed: everyone could, for example, tell a master and his servant apart by their clothes.[12] The order of things, the hierarchy of society, simply demanded distinguishable clothes. Richard Whitford emphasized that it was unsuitable for a king to wear a beggar's coat or for a wise man to wear a fool's coat.

Similarly, a monk could not wear lay clothes, because he was not a layman. The choice of clothes was not only a matter of taste, it was understood to be natural that one's clothes reflected one's character: 'the naturall dispocicion of man . . . wyll abhorre or disprayse / an vncomely / or inconuenient vesture.'[13] It follows that it was natural for a religious to wear the habit, because he had to cover his unclean body to advance virtue. This was a need in his soul, not merely an outward act of modesty. The need to wear coarse fabrics and hair shirts to irritate the skin and punish the flesh was a need that was interpreted to be beyond the control of one's mind, at least in the ideal cases that Whitford speaks about. His choice of the words 'naturall dispocicion' calls for an interpretation such as this.[14] He is seconded by William Barlow, for whom the religious habit was not at all strange as the Lutherans claimed, since it was the habit of the ancient saints.[15]

We should take a closer look now at some of the conspicuous features of the religious habit, and first return to horse hair, which was a formidable sign of piety and chastisement. Hair shirts were typically made of horse hair and worn next to the skin underneath one's clothes. They were a regular part of the attire of the Carthusians, for example, but by no means unique to the religious orders, as all devout people could wear them if they so wished. It is no wonder that wearing horse hair was a sign of true piety, since it was painful to wear. The significance of the hair shirt is well demonstrated in a legend of the murder of Thomas Becket. When his assassins found two hair shirts made with large knots in Thomas's chamber, they were filled with great fear, and returned to the dead body wondering why the earth had not already swallowed them because of their sin. Under the bishop's clothing they discovered that their victim wore a monk's habit, and next to his skin a hair shirt and breeches made of hair with knots that had eaten into the flesh.[16] It is revealing that Becket is immediately recognized by his murderers to be a holy man because of his hair shirt. The body broken by knots and eaten by worms while he was still alive was great proof of holiness, a sign of enduring pain in the manner of Christ.

The nun's veil symbolized her virginity and chastity to such an extent that if nuns were mentioned in texts, it was the veil that drew attention, and more against nuns than in their favor. Erasmus and John Skelton, for example, complained that nuns hypocritically prayed to the Virgin Mary and went on pilgrimages for her sake, but were soon ready to throw away their veils and accept money for their virginity.[17] Nuns were also criticized for liking sumptuous apparel; reformers claimed that their finery caused great pride, which of course led them into other despicable things.[18] A suspicion that not all nuns wore the habit in its simplest form is corroborated by the monk Whitford, who disapproved strongly of anyone who had fallen into the sin of *luxuria* in their habit, which was supposed to be simple and identical.[19] William Barlow, for his part, thought that friars observed this rule: 'among a thousand freers none go better appareled then an other.'[20] Confirmation of this, surprisingly, comes from Simon Fish, whom we know already for his criticism of the wealth of the Church.[21]

The tonsure was a symbol of the whole clerical order. All clerics, secular priests, monks, friars, and canons, had their crowns shaved as a mark of their humility before God.[22] It was also a sign of their special status in the world. Because the tonsure was such an important sign of the clerics' relationship to God, it is no wonder that many reformers saw it as a sign of superstition. They refused to believe that habits or tonsures would ensure anyone's salvation: Christ had indeed suffered in vain if such a sign could save a man, they argued. John Frith, for example, thought that monks were simply mad worrying about their tonsure and their hair growth, and having monthly shaves: 'What fornication thinkest thou may be compared to this transgression / if one of their shauelinges go not to the barbar / in a hole month?'[23] Keeping the new growth of hair under control required the constant attention of a barber, which could be seen as a vain interest in outward appearance in contrast to the traditional view of the tonsure as a sacred sign of humility. The tonsure certainly offered a fruitful source of merriment, and it was eagerly used. Thomas Solme, for example, mocked the clergy for freezing their heads after the first shaving of their tonsure: '[T]hey take such a cold at the fyrste shauinge that they can neuer lefte theyr hert and mend from the erth and erthly thyngis after werdis.'[24] Apparently, a permanently stiff neck was a result of the cold, and thereafter, the tonsured could see only the world, because he was unable to move his head and could no longer look up to heaven.[25]

Just as a bare crown was important for a humble monk, bare feet were a symbol of spiritual integrity as well. A religious was supposed to suffer from the cold (even if the usual cause was not the tonsure): their cells and halls were usually unheated, and only one room, called the warming room, provided warmth during the colder months.[26] As we discussed earlier, friars who wore mantles or socks could be portrayed as pursuing sinful luxury, and abbots were criticized for heating their personal quarters. It was the duty of religious to suffer more from cold than people in general, and their bare feet did earn the respect of people such as Thomas More. His admiration can be detected in his tale of a friar who met a gentleman. The gentleman asked difficult questions: Was it wise that the friar suffered the pains of cold? What if there was no hell and the friar was a mere fool for freezing his feet for nothing? It is certainly the friar who receives More's approval here: 'Ye mayster quoth the frere but what & there be hell / than ys your maystershyppe a mych more fole.'[27] For More's friar, cold feet were much more pleasant than even the thought of his feet burning in the eternal fire of hell. This illustrates how important the outward appearance, clothes, or the lack of them, was in the religious life. Being cold or hot reminded religious of the pains of the world beyond. The body was a medium between mortal life and the soul's celestial joy. Clothes were an important means of preserving, keeping, enhancing, and savoring their holiness or their attempts at holiness. The gentleman's suggesting that hell did not exist was irrelevant to the friar, since he could not entertain such as possibility.

The symbolic elements of clothes and appearance had a significant role in religious life. For religious the key elements discussed above, such as the tonsure or the veil, were messages primarily to God and only secondly to other human beings. Along with prayers, they told of the pursuit of a holy life and a better understanding of God. For someone who did not sympathize with the monastic aims, they meant only wasted effort and reliance on outward means to reach God, which must be hypocrisy and therefore failure, since Christ had not specifically asked people to wear religious habits to symbolize their piety.

8
Nourishing the Flesh

Inflicting pain on the flesh was an efficient reminder of its sinful nature, but there were other means of weakening the body as well. Food played an important role in the mystical tradition,[1] and also for early sixteenth-century religious, many of whom lived on the brink of starvation: they ate just enough to stay alive. Caroline Walker Bynum suggests that this was characteristic of women's piety,[2] but certainly abstinence from food in itself was not completely gender-specific. Although some mystics aimed to decline food and survive only with the Eucharist, death by voluntary starvation was regarded as suicide and was therefore un-acceptable. Coming to terms with eating and eating very little was far from painless. During every meal (ideally, a monastic would have only one meal a day during the winter months and two during the summer months), the body had to be reminded of the fact that sensuality lurked even in the pleasures of the sense of taste. Therefore, it was important not to feel full or satisfied: 'So that the body be somewhat punished in euery mele / and neuer to be fully saciate and contented / after sensuall desyre of appetite.'[3]

In order to understand what abstinence meant in early modern England, in a culture where not many could take food for granted, we need to take a closer look at attitudes toward food. As already mentioned, monastic rules closely regulated daily portions. For example, there were orders that did not allow eating meat at all. There were weekly fasting days (always at least Fridays and Sundays), and longer periods of purifi-cation had to be endured in association with Christian holy days, the longest fast of course being Lent. Full bellies were not tolerated since they were understood to inflame the bodies that were already feeble. Even worse, full stomachs and heavy drinking could cause 'pollution by night,'[4] i.e. involuntary ejaculation, which for chaste monks was a horror. Moreover, it was known that many a virgin had lost her chance of eternal glory as a result of excessive drinking.[5] Having such enormous power, food and drink were symbols of both good and evil. It was admitted that food was necessary – and as the Eucharist it was sacred – but its effects on the weak flesh were simultaneously cursed.

Those who had studied the classical tradition knew, for example, that Seneca had written that if one obeyed one's stomach only, one was a beast. Because bestiality was so despicable, the strictest control in all matters concerning food was essential. A dignified life simply could not include indulging in fine food. Beasts would eat as much as they could, but who, it was asked, would want to be such a repulsive creature, without self-control or reason?[6] Reflecting God's grace, the soul was to be honored even in eating. Gluttony, on the contrary, was regarded as a good way to join the troops of Antichrist.

Taking a look at popular attitudes to eating and drinking, it is practical to turn to popular literature. One of the heroes of popular stories, Robert the Devil, was from his very infancy a devilish child. He soon moved to the woods (which in itself was a sign of his inhuman ways) and led a completely evil life. Among the horrible sins Robert committed was gluttony. He ate and drank too much. He refused to respect the fast days and ate meat uninhibitedly on Emberdays and during Lent, on Fridays, and on Sundays. This was sufficient proof that he wanted nothing but to defy all authorities and even God. Gluttony was not the worst of the sins he committed, but it illustrated his inclination to evil. Robert the Devil's sins were almost unforgivable, and when he finally realized that he needed to find peace with himself and with God, as an act of penance he was made a fool. As a proof of this, he had to eat with dogs.[7] The process by which the evil Robert was brought back to humanity and humility through penance was thus linked with food. As a fool who ate with dogs for twelve years, he was placed beneath the status of man. Only through beastliness and complete purification could he return to a normal life. Robert was a layman who sinned horribly and had to pay a high price for his evil life. Had he been a cleric, his penance would hardly have been lighter. But what about clerical gluttony? Certainly it is only to be expected that any clerical character would receive treatment like Robert the Devil's for sins committed.

In the following, I will attempt to establish the role of religious in the discussion concerning food and drink. Were they symbols of gluttony or covetousness? Secondhand evidence shows that there was at least one common saying that connects greed to priests and monks: '. . . a commen prouerbe, whiche is this. He is more couetous (saye they of a couetous persone) then other preest or monke.'[8] *The images of a verye Chrysten bishop* further claimed that even unlearned people (i.e. not only the learned evengelicals) realized the sinful nature of the clergy as well as their madness and unsteadfastness, and used priests and monks as symbols of covetousness. We need not, of course, stay in the sixteenth century to find critical notions about clergy's gluttonous nature: it was a popular theme in medieval literature. For example Langland's *The Vision of Piers Plowman* introduced Wrath, a friar who ate small fish and drank weak ale during the day (which he was supposed to do), but who in the evening turned into a glutton for whom nothing was enough.[9] This tradition is apparent in the immensely popular collection *A hundred mery talys*. One of the tales is about a friar who steals a pudding from an inn on a Sunday morning before Mass. Stealing on a holy day was outrageous enough, but a friar stealing food before Mass was a fine cause for ridicule. To make his offense even more laughable, immediately after committing his crime the friar preaches about the reasons why people are not allowed to break their fast before Sunday Mass. Speaking eloquently, and gesticulating, the pudding falls out of his sleeve.[10] This, of course, gives the parishioners ample cause for merriment: the friar speaks about not breaking fast, and yet carried his own – stolen – breakfast with him. Apparently, the friar deserves no merit for not having eaten the pudding.

It is probable, however, that not all friars carried noticeable amounts of stored energy around their waists, although there is no denying that friars' appetites seem insatiable in stories like the ones discussed above. The tales were probably in part a result of the simple fact that because friars had to beg for their living and could not accept coins, they often received food as alms. This could easily lead to an image of permanently hungry friars with enormous appetites. This compels us to see the stereotypical sixteenth-century texts as a continuum of earlier satires; friars were depicted as having bellies as big barrels and filling themselves up so efficiently that 'to all goodnes they are dull.'[11] Even the highly respected Franciscan Observants would end up in the pigsty, devouring, fastidious, and looking for delicacies, beef, mutton, best drink and bread:

> Fyrst they haue befe and mutten,
> Of the chefe that maye be gotten,
>   With breade and drynke of the best:
> And that morouer so largely,
> That to farce and stuffe their belly,

> They take more then they can dyiest.
> They haue sawces with euery dyshe,
> Whyther that it be fleshe or fyshe,
>   Or els they wyll nat be content:
> To eate breade that is browne or stale,
> Eyther to drynke thynne bere or ale,
>   They counte it not conueyneñt.[12]

Ordinary food was not good enough for the characters in the satires; as shown above, they wanted meat and sauce with everything they ate, and disdained brown bread and weak ale. Furthermore, they wanted more than they could actually digest. Some radicals slandered even the Carthusians for their diet. It was said that the Carthusians had distorted the very ideal of the fast: it was not true fasting to abstain from one kind of food, especially meat, but then fill one's stomach with another, usually fish. True fasting could not include drinking heavily, either. It was not acceptable for a Charterhouse monk to fill his stomach with white bread and strong ale or even wine, when a plowman had to manage with salted, smoked bacon, barley bread, and weak ale.[13] Brown bread, of course, was a sign of poverty; the whiter and softer the bread, the better.

One text which should especially be mentioned here can be seen as serving as a bridge between the old literary tradition and contemporary pamphlets: Alexander Barclay's translation of Sebastian Brant's *Ship of Fools*, which cleverly brought fat bellies into focus. The religious stomach of friars and monks becomes the essence and the very symbol of corruption and decay of the original religion: 'Their chiefe study is their wretched / wombe to feede.'[14] The womb – which in sixteenth-century English also meant the stomach – is seen here as the focus of religious life. This idea has its source deep in the medieval tradition, and *Ship of Fools* immediately links eating too much to the question of monastic poverty. Here, again, the friars are accused of extortion from the poor, taking bread from their mouths:

> Where as they come a beggynge,
>   To the house of a poore man:
> Which hath both wyfe and children,
> And is not able to fynde them,
>   Doynge the best that euer he can.
> Yet he must vnto the fryers geue,
> Although he shulde his housholde greue,
>   Hauynge nought them selues to eate[15]

Similar images were uninhibitedly used in the religious debate from the 1520s onwards. For example, Christopher St German criticized religious people for eating too much, for having expensive tastes, and especially for the fact that some of them were more guilty of gluttony than lay people: 'some couet theyr bodyly ease and worldely welthe, in meate and drynke, and such other / more than commenly any temporal man doeth.'[16] It is worth noting here that St German wrote that *some* of the religious people coveted bodily ease. Being moderate, tactically wise, and cautious, he clearly did not want to exaggerate and present weapons to his opponents by being unreasonable. Undoubtedly there is some truth behind St German's words. Monks and nuns were usually (though not always) offered fresh, nourishing, and much better food than the poorest in society could ever have. Food was there to keep them alive and it was unavoidable that some fell for its lure. Abbots often maintained a grand lifestyle and the richest of them seem to have enjoyed a noble cuisine. The representations of gluttonous monks or friars do not, however, describe noble feasts, but rather situations in which the religious resemble pigs or other voracious beasts: it is usually quantity they were depicted to have desired, rather than quality. Gluttony in this way was a synonym for overeating.

Some critics suggested that indulgence in luxurious food was an end in itself for religious.[17] Here we should note that when criticism of using expensive foodstuffs was used, the question turned from the sin of gluttony to the sin of *luxuria*. As food had such a significant role in early sixteenth-century culture in general, the wealth of religious could easily manifest itself in food. Some critics alleged that religious could afford to have 'in his sleue a bladder, full of gynger, nutmegges or graynes: Which to make the drynke myghtye.'[18] Like white bread, wine, and strong ale, fine spices told a story of wealth. The poor could not afford to season their drink with such fine ingredients as ginger or nutmeg, which made wine a drink suitable for princes.

Some blamed religious for not practicing such abstinence from food as the desert fathers had done.[19] This would have taken the monastic diet to extremes, as some desert diets consisted only of locusts. But there were voices heard, like Erasmus's, that connected 'the delicat throote and bealy' to the worldly life of the laity. Carnal joys, good food and drink naturally include:

> deyntie dishes, the ungurgynges, the drynkynge and qauftyng one to an other / the nightly dronkennes, the bankettinges, the daunces / the gambauldes, the dissolut plaies, the knauyshe and rebaude iestinges,

which were a serious danger to the soul; they were 'lyke the laughinges of a franticke man,' and for Erasmus manifestly lay vices.[20] In a similar spirit, William Barlow wrote that the heretics raved because they wavered with every wind, and condemned fasting because they wanted to serve their own stomachs.[21] As we can see, the picture of gluttonous monks and friars is suddenly turned upside down, and the sinners are their critics. There were also voices like George Ioye's, who insisted that ordinary people's life was best equipped for holiness when it came to matters regarding food. For him, fasting was 'a temperat lyffe and perpetual abstinence from al thinge that maye cal vs from pure lyuinge and sobrenes of mynde.'[22] Good living in Ioye's opinion did not need any special effects.

This traditional image of religious as gluttons also recurs in Henry VIII's propaganda. The emphasis was on abbots who concentrated only on eating and drinking.[23] The emphasis on prelates, bishops, and abbots indicates that the ordinary monk or friar is not very interesting in this context. It must have been important to present the misdeeds of abbots, because they set the trends and morals in individual houses. When a house was dissolved, it was the abbot who gave up the seal. Attacking them was thus logical. As the dissolutions of monasteries approached, there were more and more reports on the quality of the religious life. Often the treatment of different levels of hierarchy varied: abbots would be blamed for stealing from their houses, and monks for excessive drinking. For example, the London Augustinians were accused of staying in alehouses all day and wandering into the chapel with merchants instead of actually attending services.[24] A monk of Pershore, named Richard Beerley, wrote to Thomas Cromwell applying for a license to leave his house: he felt that the company of his fellow monks was harmful to his soul because his brothers not merely refused to accept the king's supremacy, they came to services as drunk as mules.[25] The Abbot of Warden complained about similar problems. He could not control his monks, who preferred to spend their nights drinking beer to attending divine services. According to the abbot's harsh words, five of the monks were nothing but common drunkards, and he was fed up with them.[26] These complaints show us that abstinence was important in early modern mind. Seeing others drinking the night away was dangerous: drinking was feared to be a contagious condition. Naturally, a chaste lifestyle did not allow drunkenness or gormandizing or witnessing others committing these sins. This does not mean, however, that the question of sin obscured the importance of politics. Richard Beerley's letter betrays the writer's alert political eye; it seems that he wanted to be out of a house that was dangerously and stubbornly against the king in his 'great matter': the divorce.

It is interesting that English evangelicals dwelt more on complaining that monastic fasting was not genuine than on the Lutheran idea that fasting in itself was useless. It seems that food in the hands of religious, whether it was abundant or perhaps just one dry fish, was a more potent weapon in the war of convincing people of the right ways to believe than simply stating that food actually did not matter in the eyes of God if one stayed within the bounds of moderation. It seems that the attitudes of the reformers, in the end, shifted from the religious themselves and concentrated on more important, political issues.

9
Renouncing Sexuality

In a chaste life, far worse than delight in food or drunkenness was sexual lust. In fact, the concept of chastity was often reduced to apply only to abstinence from sexual acts or thoughts. The virtues of chastity were all concerned with the body and keeping it clean, but what first came to mind was sexuality. Renouncing sexuality was enhanced – and possible – only by abstaining from other sins, such as gluttony. As we have seen, gluttony was regarded as liable to lead to a restless mind easily enticed into carnality. Richard Whitford here provides us with a useful definition. Broadly defined, chastity meant controlling the whole body: 'Chastity may be called / a clene & honest state or byhauiour of al ye hole body / by ye restreint & rebatyng of ye furies & wyld hasty mocions of vice.' Chastity was a monastic vow which meant renouncing sexual acts: 'restreynt of the natural / & bodely acte of generacion.'[1] In this chapter chastity will be examined in this narrower sense, that is as a sexual matter.

It was readily admitted that of all virtues, sexual chastity was the most difficult and demanding virtue to keep. The battle for chastity was the most dangerous since it meant a continuous war against the flesh and only rarely brought victory. To those who were successful, chastity above all virtues offered the highest merit and the most glorious reward. The fight was worth fighting, not only for the wonderful prize, but also for avoiding the hair-raising punishments that awaited the fallen; the 'most horrible and depe dampnacion,'[2] the fear of which would – it was hoped – keep everyone on the right path.

In the end, everything seems to have culminated in the triangle of sex, reason, and bestiality. The Church had a profound attitude of loathing toward sexuality and especially so among its own ranks. Purging the mind of sexual thoughts was understood to be nearly impossible. Even in a monastery, where it was possible, at least in theory, to live without the sight of any women, it took dozens of years of chastisement and contemplation before sex would cease haunting a monk in his dreams. There was a causal chain in Cassian's thinking: sin led to other temptations and only enormous concentration could help the monk at night.[3] The temptation of the flesh was persistent and it certainly was beastly. Erasmus, for example, could find nothing worse than the 'moste fylthy luste of the fleshe.' To him it seemed that in sexual terms man was far worse than animals: uncontrollable and endlessly lusty.[4] The danger was formidable, the mind was weak, and even thinking about sex could play tricks on a man and lead him into temptation. Committing unnatural acts such as masturbation was a severe threat to the soul. It was against the law of nature since it wasted the seed, and was thus bestial.[5]

Marriage, the only forum for legitimate sex, was an accepted form of sexual control. Saint Jerome had even praised marriage, but only because it produced new virgins for the world.[6] Laymen could avoid bestiality by controlling their sexuality through marriage:

> Yf men sholde folowe theyr naturall dysposycyon
> Bytwene beestes and theym there were nor dyfference
> Wherfore for to haue natures operacyon
> By wedlocke it is suffred to be without offence[7]

These lines stress the importance of controlling sexuality as a means of distinguishing oneself from animals. By the sixteenth century the Church had proclaimed ideological rules about sexuality, and in theory, both having sex and abstaining from it were completely governed by the Church. In practice, of course, illicit sex was not uncommon, and sources depicting lay culture are full of frivolous and frolicking sexuality and sexual attitudes. Many devout people, nevertheless, earnestly believed that it was advisable to refrain from marriage and lived accordingly. Erasmus repeated the stereotype that marriage was a halter into which one should not put one's neck: wedlock was 'ful of moch misery.'[8] Lyndal Roper discusses the importance of marriage in lay ('middle-class') culture: in order to be respected and to gain the rights of mastership, or simply advance his career, a man had to marry. Marriage established him in society.[9] Regardless of this, marriage was not necessary for everybody. Unlike his friend Erasmus, Thomas More was married, but thought that the gift of chastity was not rare; many could pursue it and even succeed.[10] Richard Whitford wrote about the keepers of chastity – exercises that would keep the mind away from sin. The keepers that made chastity possible were prayer, abstinence, labor, decent habits, gestures, and out-ward behavior, guarding sight and tongue, and avoiding familiarity and solitude.[11] For Sir Thomas Eliot the keepers of chastity were sobriety in diet, occupation, sharpness of the inner apparel, refraining of the senses, minimized and honest communication.[12] Whitford stressed that these keepers could best be practiced in religious houses where strict enclosure within the monastery deprived the religious of the opportunity to commit most sexual sins.[13] Because delight and comfort did not go together with chastity, the ideal monastery offered austere surroundings to enhance the climate of chastity.[14]

Sexual chastity had various forms, and we will attempt to explore the patterns of renunciation. The original theological idea behind the loath-someness of everything sexual was basically quite uncomplicated. Sexu-ality was fervently hated by so many churchmen because it was a result of original sin. Sexuality in itself was a punishment, not a gift, for human-kind. Christ, who in the Christian tradition was seen as asexual, had set an example for others to follow so that they, too, could achieve salvation by counteracting the curse of sexuality brought upon humanity by Adam and Eve. Virginity was Christ's answer: to renounce sexuality was to laugh in the face of eternal death. This is why the strictest views gave little credit even to controlled sexuality; it was merely tolerated (as we saw in the case of St Jerome). In the end, remaining virginal in this confusing world of sexuality was always the best option in the eyes of the Church.[15]

Early sixteenth-century English people well understood the historical meaning of female virginity and could see its many virtues. Thomas More, for example, wrote about the political importance of the Vestal Virgins in Rome and of the fact that the fortunes of the state depended on their virginity.[16] One need not, however, go that far in history to find that the power of a virgin could be frightening, as in the case of Joan of Arc, whose virginity Marina Warner interprets to have been '[j]uxta-posed to the vivisected and dismembered body of the kingdom, her virginity provided an urgent symbol of integrity. By synechdoche, Joan's intact sexuality stood for the whole of her and, in the ambitions of her supporters, for the whole of France.'[17] Virginity was a special state, deserving great admiration, and in extreme cases generated a cult. Vir-ginity in many ways meant freedom from the pressures of society.[18]

Medieval women's virginity could be divided into two categories: on the one hand, there was spiritual virginity – Bottomley calls this sacra-mental view of virginity, which was not centered on the hymen at all. This was virginity that was more a mental state than a physical one. On the other hand, lay virginity was mainly a physical state, and the preservation of the hymen played a major role in this type. This was the kind of virginity that had great value in the marriage market of the social elite. For Bottomley, this is utilitarian virginity.[19]

In late medieval discourse virginity was an exclusively female issue. In everyday terms chastity in itself was a gendered virtue. Just as obedience was regarded very much as a female virtue in society in general, chastity and virginity similarly became women's domain, whereas honor was the major identity attribute of men.[20] Virginity was a field in which women especially could and were expected to excel, but, simultaneously, they were marginalized because of the gender specificity of this ideal.

Even though a virginal life was not a guarantee of a pure life, it was at least as close to it as was thought possible. Virgins deserved a special place in this life and in the next. Even in heaven there were different categories for the blessed: virgins would receive a hundredfold reward for a clean life, whereas chastity in widowhood brought a sixtyfold reward, and cleanness in marriage brought a thirtyfold merit. Virgins were also rewarded otherwise: they were allowed to sing the Virgin Mary's holy song and to sit close to the angels and Christ and follow him everywhere.[21] Allowing all this for women, late medieval religious culture made a great concession for these creatures who were otherwise often perceived as spiritually weaker than men.

The culture which hierarchized everything did so even with virginity. There were three estates of the flesh: marriage, chastity, and virginity.[22] In other classifications, chastity could be divided into different categories when the first and most valued degree of chastity was the virginity of those who vowed chastity and lived in religious houses, the second was the virginity of maidens living in the world, and the third chaste widows.[23] It was, indeed, usual to think of widows as a specific group who, if they remained chaste for a long time or even professed as nuns, would eventually regain their virginity, if not physically at least spiritually. A good example of this was St Bridget of Sweden. The existence of this category shows that late medieval people often had an open mind, and that black and white could be supplemented by shades of gray. This does not mean that those who had lost their virginity were not stained in some way; great effort was needed to become white again. This process was compared to the making of linen: when flax grew it was green, and it needed drying, beating, watering, and keeping in the sun to make it white. Similarly, chastity was hard work and it paid never to fall.[24] After a period of discipline and abstinence, virtue was natural to those who had not erred, but credit was given to those who fought their way out of the darkness after their lapse. Even fallen women could achieve and excel in virtue, and perhaps surpass the virtue of 'genuine' virgins. St Peter of Luxembourg stressed that the loss of virginity did not mean all chance of a sanctified life was lost. He offered Mary Magdalene and St Mary of Egypt as examples of women who had fallen but who in the end were more virtuous than they ever could have been had they remained virgins: 'them adourned so wel and of so noble vertues so rychely that they were more worthy after than before.'[25] Experience could, curiously enough, purify.

This open-mindedness did not extend everywhere. Virginity was an ideal that followed no equality patterns. Guido Ruggiero has shown that in Renaissance Italy virginity too was an issue of class. Virginity was especially important for unmarried women of the upper classes, but its importance decreased as you went down the social scale.[26] Similarly in English jestbooks, a servant girl could be made to lose her virginity to save her mistress's reputation without anyone blinking an eye, whereas a noble woman's virtue would attract endless admiration. An example which must have amused contemporaries is given in *The deceyte of women*, a collection of tales about licentious women. In one of these tales, an adulterous woman decided to cheat on her husband, and puts her maid in bed with her husband, thinking that he would not notice the difference. While she herself enjoyed the company of her lover, the poor servant girl lost her virginity. Out of fear and out of the need to save 'her maystres honestie' the servant dared not protest. Needless to say, the husband spotted the difference between the two women immediately. Neither the girl's own wishes nor keeping her virginity intact meant anything to him (nor did it to his wife), since he happily relished the situation afterwards: 'Mary for an olde hore, to haue a yong proper mayd.'[27]

A servant's virginity was irrelevant, but noble virgins were white flowers, lilies, pure and innocent. Chastity could be depicted as the beauty of a king's daughter.[28] For a noble woman more inclined to holiness than the lady in the example above, it was sometimes possible to preserve herself for the exclusive use of Christ. The following example is once again St Werburga:

> My purpose was neuer / maryed for to be
> A lorde I haue chosen / redemer of mankynde
> Ihcsu the seconde persone in trynyte
> To be my spouse / to whome my vyrgynyte
> I haue depely vowed / endurynge all my lyfe
> His seruaunt to be / true spouses and wyfe.[29]

St Werburga is idealized, lifted above ordinary humanity and ordinary womanhood. She is Christ's own. The adjectives used of Werburga depict the ideal female heroine: she is 'obedyent', 'adde', 'demure', 'stable', 'sobre of wordes', 'humble', 'meke', and 'mylde'.[30] The fact that she is royal only emphasizes the strength of her decision and her virtue in defying her father and leaving the world.

The extract from St Werburga's life shows an important element in monastic thinking: she chose to be Christ's spouse. The concept of *sponsa Christi* had earlier been applied to both sexes but in late medieval understanding, only women were depicted as Christ's brides.[31] The very idea of monastics being Christ's betrothed originates in Corinthians 11:2: 'For I betrothed you to Christ to present you as a pure bride to her one husband.'

An aspect of virginity that applied also to monks was the angelic life, *vita angelica*. It was essential for understanding medieval virginity, since it was by remaining virginal that angelic life was pursued. Some theologians even considered that in God's eyes a cloistered life was more valuable than the life of angels in Heaven, because angels did not have to strive for their virginal and holy state, unlike monks, who had to watch themselves both day and night.[32] The fact that monks had to do so gives reason to suspect that, in contrast to what Penelope Johnson has suggested, women religious were not necessarily regarded as holier than men in the eyes of their contemporaries. Johnson posits that when a woman gave up family life, in the eyes of others she sacrificed more than a monk did.[33]

Looking from another angle, male virginity could be especially respected because when men wished to remain virgins (or chaste), they had to strive against their flesh in very concrete ways. The danger of nocturnal ejaculations was a constant reminder to monks that virginity – and chastity – was especially difficult to preserve. The battle against the flesh could, at its best, refine the monk and show him the way to God as well as lift him above the angels. Patricia Crawford points out that early modern 'medical writers regarded the male body as the norm. Since men produced more seed, their sexual urges were more powerful. From this, one lawyer extrapolated the conclusion that men were by nature less able to be chaste.'[34] It should be noted, of course, that here lies another of the history's controversies: some writers thought that it was women who were more lustful and more uncontrollable than men. As Natalie Davis notes, 'clerical celibacy for the superior sex had been thought a real possibility whereas for the female it had appeared an exceptional achievement.'[35] Both views are correct; often it seems, however, that whichever view was proposed, it was done to enhance male superiority. Remaining virginal or chaste in itself never guaranteed a seat next to Christ in Heaven. The favor of God was more than easy to lose. Falling from grace was no less complicated than having a lustful thought; even feeling pride in being a virgin overrode the virtue,[36] and so did talking about sin. Mirk's *Festyuall* includes an exemplum of an abbess who is known for her virtue, but who to everyone's surprise appears to her nuns after her death warning them against falling into the same trap as she had: she has seriously endangered her immortal soul because she has talked so much about sin:

> / ye knowe well that I was a clene mayden as for ony dede / but I had grete lust to speke of synne.... By this may ye se what peryll it is to speke ydle wordes & harlottry speche / wherfore this same pystle sayth thus (Abstinetis vos a fornicatione).[37]

Even small words could mean falling from the true path. Words were thoughts and thinking of sinning was sin. Resisting and being able to resist sexual temptation brought merit to religious. It should be remembered that people who were not capable of sexual acts such as castrated men were not fit for the angelic life. Since a eunuch was not capable of committing sins of the flesh, his chastity was not regarded as meritorious as the chastity of those capable of this sin.[38] A eunuch did not suffer from the lures of the flesh and thus his striving was not as great an effort as the striving of those who were capable of carnal acts and sensuous feelings in their flesh: 'for therin wolde be loste the meryte that good men haue in resystynge of the deuyll / and the refraynynge of theyr fleshly mocyon.'[39] Monasteries were places for the voluntarily impotent, just as they were for the voluntarily poor. Just as the body was an instrument, sexuality and sexual organs too could be instruments not only to sin but to salvation as well.

It seems, however, that in popular sentiment monks *were* eunuchs, possessing completely useless genitals not by castration but by choice. Men without testicles were rarely regarded as fit for more than a religious life, or at least a eunuch probably felt it would best serve his purposes in life. Another tale from *The deceyte of women* reinforces this notion: a young man finds an ingenious way of deceiving his lover's husband so that he can spend more time with her. The young man covers his private parts with a sheepskin and deceives the husband into thinking that he has been castrated. He complains greatly about his situation and pleads that the husband would help him to get into a monastery 'where that I may haue my meate & drynke and that I may serue god for I am not mete to be in the worlde.' The husband takes pity on the poor man and takes him as his companion to escort his wife even on pilgrimages. The deception is complete: the adulterous couple are now free to satisfy their lust without the danger of the husband prying on them.[40] In point of fact, this tale tells more about the worldly ideal of a perfect man: he has to be in perfect condition to perform his duties, and if he is not, he is completely inadequate. In this tale being inadequate means that the young man should leave the world to serve God – as a man without testicles, there is nothing for him in this world. Sexuality plays an important role in society, and this story thus further emphasizes the notion of the asexuality and marginality of religious.

Sexuality was an essential part of everyday life both in and out of monasteries. For ordinary people, sexuality presented somewhat different problems. Not everyone cared about perpetual virginity, and some people saw marriage as surpassing virginity in virtue and in value. The flesh was nevertheless difficult to live with. In religious houses, medieval traditions were still alive before the dissolutions, and great effort was exerted in proving their success and excelling in virtue. The marginality of religious was not all negative: through excellence they could turn the marginality to work for them, not against them. Even though they were a relatively small group, they could easily dictate ideals for the majority, with the authority which God's word brought.

It is now time to take a look at what was thought of the sexuality of the religious. Groups of men or women living in communities, in secluded conditions, unchecked and uncontrolled, gave ample possibilities for imaginative minds to create images, gossip and tales of the wildest kind, whether based on facts or not. As a result, religious could be presented as the greatest of all adulterers or seen as so strict in their religion that they could not even read or discuss matters such as carnal love.[41]

Whatever the images, one rule could not be disputed: for a monk, the flesh was dangerous, both in thought and in practice. The sinful acts include adultery of all kinds: masturbation, lecherous gestures, and incest, which is especially mentioned because of its horrible nature. One should

> auoyde and eschewe all auoutry of weddyd folkys / and all formacion of syngell folkes / all wylfull pollycions / inordynat Lecherous gestures: and specially synne agaynst kynde whiche cryeth vengeaunce before god.[42]

Lechery was further hierarchized into several categories. From the less sinful offense to the most sinful, the list reads: fornication between a single man and a single woman, adultery, unlawful deflowering of virgins, incest, and finally sodomy, sin against nature: for example, a man or a woman with a beast, a man with another man or a woman with another woman.[43] Especially for religious, thinking about lechery was a great sin and from there, the ladder of sexual sin was seen as an easy one to climb – or descend. Simply touching the flesh could lead instantly to the actual sexual act or masturbation.[44]

The chastity of religious was not only of great concern for religious themselves; many ordinary people wanted to think of them as an example of an ideally chaste life. On the other hand, in many texts monks, friars, and nuns emerge as emblems of carnality and fornication.[45] Richard Morison, for example, wrote that genuinely spiritual people lived *outside* monasteries. Monks, in contrast, were lecherous and unashamedly used other men's women for their lustful purposes.[46] Morison implies that God actually preferred villages and towns, and the lives of ordinary people, to the world of the cloister which to Morison seems to be part of an artificial recreation of a religiosity long extinct.

Morison's words were nevertheless moderate compared to many other insinuations about the lack of morals in religious houses. The boldest texts accused religious of systematically committing sodomy or at least covering the tracks of anyone doing so. Homosexuality,[47] however, was not the worst of sodomitic practices: in St Thomas Aquinas's list, the most horrid was bestiality, homosexual practices coming only second. The third was intercourse with someone of the opposite sex in a position that was not approved of by the Church.[48] *Regularis Concordia* put homosexuality and paedophilia before heterosexual temptation.[49]

Sexual sins were taken seriously and it is no wonder that the evangelicals used harsh words to describe failures in chastity: 'Sodomites,' 'Gomorreans,' and 'monstrouse adouterers.' James Sawtry wrote that their horrible sins could not even be discussed in writing because their monstrosity would poison the paper, the paper would infect the readers' breath and then corrupt the air and the ears of anyone who happened to hear the words read.[50] Sodomy, or homosexual acts, indeed was a sin of silence that authors found extremely difficult to discuss – it was felt that even mentioning the sin would lead people into sodomitical practices.[51]

Evangelicals could say, for instance, that monasteries were the creations of the Devil, or gifts from him to all who desired to dwell in hypocrisy and sodomy. In the following extract, the Devil himself speaks and confirms that he has given the monasteries as a gift to the

world so that all sins could be easily committed. Sinning could begin
with idolatry and sodomy:

> And to hyde all thynges from the ryght of the people wee have geven
> you large monasteries to take your pleasure in. In the whych solitary
> places under the colour of chastitye, ye may live in shamefyll lawdrye
> [idolatry] and commit most abhominable sodomites.[52]

The fallen angel delighted in offering people opportunities to sin. For
this purpose, religious houses were ideal. Perhaps it is not too far-fetched
to return to the marginalization of monks here. As Alan Bray proposes, it
was unmanly to be a sodomite.[53] Because monks' chosen way of life
made them different, it was quite logical to see their gender role as
blurred. This easily led to thinking that in the cloisters the unmanly
sin flourished. Because most religious houses were enclosed communities, anything could go on behind the walls. It was as if the walls
that kept religious out of view inflamed the imagination. Everything
that differed from the norm (it could be called the 'Other') was intri-
guing and frightening. The marginalized monks and nuns were out of
sight and therefore strange. Anything was possible.

Because the evangelicals favored the solution offered in the Bible to
marry in order not to burn, they disapproved strongly of religious who
sinned. They were certain that it was not humanly possible to lead a
chaste *and* unmarried life – it follows that no monk, nun, or friar could
be chaste. For them, chastity was a 'gyft from god farre aboue the power
and strength of nature,' but they did not believe that God had given
religious the strength to remain chaste. Therefore, in monasteries they

> do se them bothe men and women, whiche beyng letted and holden
> in within the prysones of monasteryes, & with iyerne [iron] lockes
> and grates can not haue the company of man or woman, to synne
> more fylthelye & more abhomynablie (sauyng reuerence) in sodomy-
> trie or buggerie, then a man dare be bolde to vtter or expresse. For
> albeit they do neuer so moche conteyne outwardly, as concernynge
> the bodely acte: yet that nothwithstandyng they do inwardly boyl &
> brenne with vncleane luste.[54]

The author seems to think that it is more difficult to live chastely in a
religious house than outside it. No monastic prison, no iron locks could
change that. The author refers to sodomy and buggery, which here
certainly mean homosexual acts; he is horrified beyond words and

further enforces the notion of sodomy as a sin that could not easily be discussed. The author admits that religious do not always commit the act, but that they could sometimes commit it in thought, which is just as disgusting. The issue of homosexuality emerges in the debate from time to time. It is sometimes noted by the reformers that the whole clerical order was inclined to have 'Roman habits': according to Englishmen, homosexuality had been widespread in both ancient Rome and in the Rome of the popes.[55] Richard Morison also insinuates that monasteries are nests of that vice:

> I would scarce believe that men could teach nature a new way, except it had been proved to their teeth and uttered by their own selves. They that be learned know what I mean, and what they are. Paul laid the same fault to the Romans. They that be unlearned will much marvel (except they have been brought up with monks and friars) how young novices may stand instead of young wives. I have said enough. It stinketh too sore to be stirred too much.[56]

Morison here hints that homosexual activities occur in religious houses: the law of nature is ignored, Roman habits are employed, and novices stand for wives; all these are clues to what he suspects are sodomitical acts between men. Morison finds these practices revolting, and being politically astute, he finds condemning them a means to flatter the king. Quite cleverly he emphasizes the god-like qualities of Henry, and praises the king for his mercifulness because he had not instantly executed all these sodomites. Morison confesses that he himself feels that all sodomites should suffer a horrible death.[57] Morison's criticism here turns into a propagandistic trick, which cannot go unnoticed.

Even though the example above raises no doubt about what kind of sin is meant, reading the pamphlets usually requires caution on the question of sodomy, because reading the term sodomy as homosexuality is far too easy, and in many cases quite misleading. Sometimes the texts refer to Sodom and Gomorrah, and compare religious houses to them, but not in all cases as centers of homosexuality, often simply as nests of sexual vice in general. For some, the complete destruction of Sodom and Gomorrah was as child's play compared to religious houses, since in them innocent souls were being lost every minute; the end of the two cities, in contrast, had been instantaneous and therefore merciful.

An illustrative example of the symbolic Sodom and Gomorrah of the monastic order can be found in *Reade me frynde and be not wrothe*. The author talks of the lusty monks who teach the young their horrible habits, including their unnatural acts, which are much more varied than homosexuality alone:

> Their cloysters are the devyls mewes,
> Farre worse then any stewes,
> Or common places of whordome:
> They are the dens of bawdynes,
> And fornace of all lecherousnes,
> Lyke vnto Gomer and Sodome,
> Yonge laddes and babes innocent,
> They brynge in by their intysement,
> To their lewde congregacyon.[58]

Even the Observant Franciscans are said to have lived in sin (though they are noted not be guilty of open adultery): although they pretend to be chaste, they have 'pollucyons detestable' and 'inwarde brennynges intollerable.'[59] These allegations show that what was in the margins was easy to denounce as dangerous and sinful. It is probably against the expectations of many, but sodomy did not actually draw the most attention when the sexual sins of religious were under scrutiny. It was adultery which raised most interest and it is not surprising that it did so because evangelicals strongly favored clerical marriage. Simon Fish stated that there were a hundred thousand idle whores spreading diseases in the country because of the clergy. No man could be certain about his wife or his daughter. Even worse – no man could be assured that the bastards of the tonsured class would not inherit his property:

> These be they that haue made an hundreth thousand ydell hores yn your realme whiche wolde haue gotten theyre lyuing honestly, yn the swete of theyre faces, had not theyre superfluous rychesse illected theym to vnclene lust and ydelnesse. These be they that corrupt the hole generation of mankind yn your realme; that catche the pokkes of one woman, and bere theym to an other; that be brent wyth one woman, and bere it to an other; that catche the lepry of one woman and bere it to an other; ye, some of theym shall bost emong his felawes, that he hath medled with an hundreth wymen.[60]

That the monasteries acted as a means of spreading, for example, the lethal syphilis was a formidable accusation, so frightening was the disease. And, of course, the other plague for Fish is the issue of women's weak nature and fickle disposition. Which woman will work days and nights for threepence a day if a friar, a monk, or a priest pays her twenty pence for one hour? he asked.[61]

The image of whoring friars especially was strong; it even emerges in the pages of Thomas More's *Dialogue against heresies*, though this time in a reversed form. More seems to be commenting on a literary tradition as well as rebuking people for unfairness toward friars or the clergy more generally. More asks his opponent the following question: Several young women accuse a friar (the women's confessor) and say that he had promised to ease their penance if they slept with him. Who should one rather believe, the young women 'not very specyally knowen for good' or the friar who has a good reputation? More's (imaginary) opponent unhesitatingly admits to believing the women since 'ye thynge is so lykely of it selfe / that a freer wyll be womanyshe / loke the holy horesone neuer so sayntly.'[62] Regardless of the opponent's answer, however, the friar is innocent and the women deceitful. More regrets that people were more willing to believe two simple women than ten good men on these occasions.[63] More's personal sympathies are on the friar's side; he is not willing to think the worst of religious.

Although friars were a popular subject in satires, they were not always on the receiving end. Nuns, too, could easily be branded as wanton – we have already seen how they were seen to leave their veils for any call to sin – and nunneries brothels, even to such an extent that brothels were mockingly called nunneries. This was a popular literary image,[64] and it readily insinuates that it was self-evident to everyone that all nuns were whorish.[65] This kind of comparison was easily made, as the parallels between nuns and whores were plentiful: both prostitutes and nuns formed communities, both lived in groups with a woman as their head, and both nuns and whores had strict rules in their houses. This is at least what the contemporary understanding was like. Women living in a community always made a suspect group, more so if they were marginalized among women as nuns and even more so, prostitutes. It did not help much if there was a man to watch over the community of nuns, since the nuns would certainly seduce their priest to destroy the virtue of a nunnery:

> It is a perrellous poynt for Nonnes chastite to be reclused in siche a cloister where Preistes be to familiare and bere all the rule beinge at meall tyde bedde and borde within the place.[66]

Evangelicals simply thought that the proper way to tame a lusty nun was to marry her and make her bear children:

> The Nunnes or sacred vyrgyns now in our tyme, are for the most parte
> lustye damoyselles, and of florysshynge age / and created of god, to
> the entent that they shuld wedde and brynge forth chyldren.[67]

For the reformers a woman's place was in a family household, in which
she would take care of her husband, and have his children. In a sense, if
a nun did that, she gave up what had been sacred in her: the ability to
come close to Christ through her virginity or her pious lifestyle. It is
interesting that the above author talks about sacred virgins and nuns.
Perhaps he even believed that virgins *were* sacred or that at least they
had been so in the past. It is possible that the author's choice of words
indicates that he truly regarded the state of virginity as sacred and holy
in the eyes of God. It is even possible to read in his words that now the
place of the sacredness must be taken by a more worldly activity in the
household, but that sacredness could be retained in another form, as a
mother.[68] It could be cynically wondered why the same women who
had earlier been blamed for whorishness were now fit to be mothers.

It was, as said earlier, in the reformers' program to sanctify marriage
above virginity, and when it came to the question of women's role in the
world, it served the reformers well to promote motherhood. This process
is visible anywhere the Lutheran influence reached. For example,
Sweden issued a new church order in 1571 which strongly promoted
the role of the mother as the correct female ideal and the home her
proper sphere.[69] Merry Wiesner sensibly proposes, however, that
women did not necessarily accept this in their lives.[70] If it was so, it
becomes conceivable that traditional values – especially of virginity –
lingered much longer than has often been accepted.[71] In her *City
Women and Religious Change* Natalie Davis argues, however, that not all
women needed to protest against the new ideal of marriage, and notes
that, for example, the ideals favored by the Protestants (for example, of
companionship in marriage) were not their inventions, but rather already used by the Catholics.[72]

Simon Fish bluntly stated in 1529 that the king should send monks
back into the world so that they could marry (instead of using other
men's wives), work for their living, and set other lazy people a good
example. This would result only in good: the number of thieves, prostitutes, robbers, and idle persons would significantly diminish.[73] The
exclusive nature of celibacy could be and was used as a weapon by the
reformers. They recalled that Christ himself had said that he did not
expect celibacy of everyone, and the reformers now blamed religious for
taking a vow which they could never keep. The reformers were con-

vinced that total sexual abstinence was the virtue of a very small minority of people and that religious certainly could not achieve this goal. Such a message was a one-way street that offered religious no means of defense or escape; for if the religious claimed that they *could* stay chaste, they, from their critics' point of view, made Christ a liar, because Christ had said that chastity was 'a gyfte synguler.'[74]

The question of clerical marriage put matrimony, which in popular literature so often was unashamedly scorned, into a very favorable light. It was useful for evangelicals to idealize marriage as a counterpart to the ideals of chastity in the clerical orders. William Walter stresses that there are virtues in love and in begetting children and that the union of a family is a strong one:

> Better than chaste loue / what thynge is to be loued
> Whiche is grounded in holynes and also in honeste
> Frendshyp and affynyte is therby encreased
> In one body togyder ioyned be
> Chyldren borne in wedlocke be lawfull and fre
> They be combyned with bonde so charytable
> That nothynge but dethe can make them separable.[75]

Those who supported the religious did not speak seriously against the virtue of marriage; on the contrary, they too emphasized that chastity was a virtue meant for a special group, which was small in number. They admitted that not all clergy were able to stay chaste, and because the accused admitted this, their critics would attest that the Church had created a deliberate system to allow sinful living for those who had sworn chastity:

> The Pope sayeth all monkes / fryers / and nonnes shall vowe and swere chastite be it geuen them or not / my prestes also shall not be wedded / but as for to kepe hores and rauyssh other mennes / doughters and wyues / shalbe despensed with all.[76]

Likewise, John Bale condemned religious, priests, and beguines[77] for fornication, for acting unnaturally, and for not supporting the institution of marriage. Bale thought that they thus considered adultery the holiest state. They were 'spyrytuall Sodomytes' and the world would soon see how devilish they were, holding marriage in contempt and living in horrible filthiness. Bale noted that there was no color that could 'hyde all their knaueryes, as what that counterfett chastyte of theirs.'[78] It is clear that the reformers wanted to reorganize the entire question of sexuality in the Church.

Bale's judgment was not shared by everyone. Some years before his time, a strong proponent of monastic virginity and chastity, Bishop John Fisher, spoke about the importance of inward virginity, of which such great examples as Christ, the Virgin Mary, St John, and St Paul had given a living and a glorious example. As an answer to the Lutherans who had attacked and persecuted religious, especially in Germany, Fisher wrote that there indeed were thousands of genuinely chaste monks and nuns in Christendom: 'it is nat to be doubted, but in all Christendome be lefte many thousandes, whiche at this houre lyue chaste, and truely kepe theyr virginite vnto Christe.' Fisher stressed that it was the Lutherans who were wrong: *they* had dismissed the ideal of virginity for the wrong reasons, and this had then resulted in neglecting religion itself. Their erroneous reasoning had resulted in German monks and nuns returning to the lusts of the flesh. Bishop Fisher is especially unforgiving:

> Nowe, let vs se, whether ye sede of gode worke this high frute amonge the Lutherans or nat. No. no. nothyng lesse... The religious men forsake their religion, and retourne vnto ye world, and take them queanes. The virgins that were consecrate vnto god, & had promysed to kepe them selfe as true spouse vnto Christe, nowe gyue their bodies tyll all wretched pleasure, and suffre them selfe to be stuprate and abhomynably defyled and soused in all carnalite.[79]

The actions and speeches of the Lutherans brought eternal destruction to many. It was they who were 'abhomynably defyled and soused in all carnalite.' Fisher's anger was not merely directed against the heretics in general; the greatest enemy was Luther himself, whose personal life was under his judgemental eye: Fisher found it abominable, and could not decide which of Luther's atrocities was the most foul: marrying a nun, degrading the sacrament of marriage, his adulterous marriage directed against Christ, or the fact that Catherine von Bora gave birth to their child just six weeks after their marriage. Of Luther's last mentioned sin Fisher only noted sarcastically: 'This was spedy worke.'[80] In Fisher's eyes, Catherine von Bora had been *sponsa Christi*, consecrated as Christ's future bride; by marrying this woman Luther had expressed great contempt for Christ. This was much more than fornication. Clearly, Fisher still thought that the vocation of a nun separated her from the sphere of ordinary women; if the nun fulfilled the expectations of God, she would enter his secret chambers and serve him in eternal glory. For Fisher, virginity was still consecrated to God.

Just like Fisher, Thomas More fiercely attacked those who wished to abolish clerical celibacy. On several occasions, he returned to the horror he felt at the marriages of Lutheran ex-monks, friars, and nuns. So often did he refer to what he saw as incestuous marriages that he must have been obsessed by them.[81] He found the very idea abominable and stressed that clerical marriage remained sinful whatever the heretics said. Those members of the clergy who married committed, in his opinion, 'incestuouse sacrylege and very bestely bychery.'[82] For More, not unlike Fisher, the horror of Luther's marriage was centered on the fact that both partners had taken a vow of chastity and especially on the fact that as a nun, Catherine von Bora was a bride of Christ. Luther committed a sin greater than just breaking his vows to God: he committed incest, since by marrying Christ's bride Luther married his own sister and simultaneously committed sacrilege, since he made Christ a cuckold by marrying his wife.[83] In this, More was in complete agreement with Fisher. More spared no words when he described Luther: he was 'a fonde frere,' 'an apostate,' 'open incestuouse lechour,' 'a playne lymme of the deuyll and a manyfest messenger of hell.'[84] Luther was weak since he let his lust guide his path, and his lust now made him an advocate of fornication.[85] More was outraged by Luther and others who followed his example. Lutherans had lost all credibility in his eyes, and More scourges them with his words:

> Who would not confound heaven and earth, sea and sky when he sees a Lutheran bishop – a man who has broken his vow, shattered his faith, violated the chastity of his priesthood, who wallows in continual incest, which he prefers to call marriage, who shakes his ass as he preaches about virtue – suddenly pontificate about his grave and weighty rules and regulations concerning the worship of God as if he were sent down to us from heaven.[86]

More could find no reason whatsoever to justify clerical marriage.[87] To him, breaking the vow of celibacy meant the destruction of all religion; furthermore, the loss of the ideal of virginity and chastity was a sign of the approaching destruction of the world. More, like Fisher, found a fitting example of this horror in Lutheran Germany. He was certain that religious houses would be turned into brothels after the religious have left them.[88] More protested strongly against the idea – for him a heretical one – that clerical marriages (of priests or of religious) would reduce the number of prostitutes and procurers. On the contrary, these incestuous and horrible marriages would make them all 'starke harlottys,' and all those who help them to marry nothing but 'stark bawdys.'[89] The opinion of Fisher and More, which was in line with the 'official' opinion of the time, as Henry VIII did not accept clerical marriage either, did not make the issue go away, and after their deaths the debate on clerical marriage continued. Among others, Thomas Starkey tried to influence the king by the very typical reasoning that allowing priests and monks to marry would make them chaste. Moreover, clerical marriages would enrich the land because more children would be born. This certainly would be an outcome which any reasonable king would appreciate.[90]

Now that we have discussed literary views on the chastity of monks, nuns, and friars, it will next be asked what was thought of the chastity of the English religious when the religious houses were attacked immediately before the dissolutions and during the process of dissolving the houses. Perhaps the most influential person over the fate of the religious was the vicegerent Thomas Cromwell. In 1536 he gave injunctions to the clergy which are interesting if we try to trace Cromwell's own ideas on the clergy. He ordered the clergy to mend their ways: the clergy were not allowed to exhort anyone to go on pilgrimages or to pray for the saints, they had to donate all funds reserved for images to the poor, and they had to abstain from alehouses and taverns. Drinking and gambling were forbidden, and all in all, it was stated that the duty of the clergy was to set a good example to the people.[91] It is interesting that the question of chastity is not made an issue here, except insofar as women were banned from male religious houses whatever their role. First, this could mean that virginity and chastity as such did not stand out as especially important qualities or virtues in Cromwell's thinking. Second, it could imply that there were differences of opinions between Henry and Cromwell. Henry wanted clerical celibacy; Cromwell perhaps had his doubts. A third explanation is also possible: that on this issue Cromwell was satisfied with the standards of the clerical estate. Whatever his reasons, a strong condemnation of the sexual misbehavior of the clergy should have been expected if he was certain that the religious (or other clergy) were living in sin.

Henry and Thomas Cromwell's policies tactically were well thought out. Visitors were sent out to the religious houses and, inevitably, the reports almost invariably found some – some very old, some brand new – evidence of misbehavior in most houses. It did not count if a sinful act had happened only once, years ago – it was still a valid proof of decay. Naturally, it is impossible (and beyond the scope of this study) to draw any conclusions about the standards of morality in the religious houses from these reports. What can be said is that the reports were eagerly used. Henry had no difficulty in passing the Act to dissolve the smaller houses after the contents of the reports were made known. If the reports are critically scrutinized, it can be seen that the crimes committed were far from the worst forms of sodomy in their seriousness. For example, the Abbot of Bury was found otherwise impeccable, but guilty of enjoying dicing and card games.[92] It could also happen that no sins or misbehavior were found in a house, but the commissioners expressed in their report a strong suspicion that there certainly had been some problems, which they simply could not unearth.[93]

The propagandistic images could be used to change opinions for political purposes; some of the commissioners' reports sound utterly damning in their horridness, and it must have been easy to believe that they held at least a grain of truth. Naming the culprits was especially effective. No one had ever claimed that all religious were saints, but the commissioners now claimed that some of them almost exceeded in evil anything that could be imagined. For example, they reported that the Abbot of Cerne kept concubines in his cellar and allowed them to sit with him at the table. This was not enough: he allegedly spent the monastery property on these women and on the marriages of his children.[94] The Abbot of Fountains was accused of wasting wood, stealing, and keeping six whores.[95] A prior was found in bed in his London convent at 'aboght XI of the clock in the fornone upon a fryday' with his whore, as John Bartelet put it. He told Cromwell that the prior had offered him £30 to keep the incident a secret.[96] In Langdon, Dr Layton ended up breaking the abbot's door as the abbot was in his chambers (allegedly) with a woman. The unfortunate abbot was taken to Canterbury and put in prison.[97] Sometimes the religious themselves did the reporting. The monks of Warden were, according to their abbot, guilty of taking whores from a brothel to the monastery vineyard.[98]

These cases show the intensity with which the visitors operated. Intruding into religious life, they certainly aroused anger and aggressive behavior. For some religious, refusing to cooperate with the king's men was a natural solution. When publicized, however, such incidents rebounded against the religious life, though it is very difficult to say what their far-reaching consequences were. Did the laity even hear about these cases? If they did, did they have any effect on their attitudes towards religious people? The answer to the first question must be yes; if they were heard in Parliament, why should the rumors not have spread? Regarding the second question we should note that as a part of a visible smear campaign, rumors such as these may even have worked against their publicists and aroused sympathy for religious; this is mere speculation on my part, however. It must be remembered, though, that the cases mentioned represent only a few disorderly religious houses – there were many that had experienced no troubles in years and people might have been aware of that. The reactions in the North, where people took up arms, do show that it was not an easy matter to convince people of something they did not believe.

The commissioners' reports make especially interesting reading when they talk about nunneries. Perhaps surprisingly, it could be concluded that the morals of the nuns were more praised than reproached. It is certainly possible that nuns were more virtuous than monks, but if that point is set aside, the reasons for this could be found in the politics of the suppressions. Women politically[99] were not as dangerous as monks and it was thus possible to praise them, as the commissioners praised the nuns of Catesby:

> Howse of Catesby we ffounde in verry perfett order, the priores a sure, wyse, discrete, and very religyous woman, with IX nunnys under her obedyence as religioius and devoute and with as good obedyence as we have in tyme past seen or belyke shall see.[100]

Occasionally, monks were praised as well. In Kent the commissioners reported having found two virtuous houses which were both clean and well maintained and whose inhabitants were well liked by their neighbors.[101] It must be added, though, that these two houses had just surrendered their seals to the king, so there was no longer any need to condemn them. This speaks about the nature of the king's campaign and about the fact that the conclusions in the reports were more dependent on the political situation than the actual virtue in each house. If a religious house was loyal to Henry, it was probably much easier to commend it than one that categorically refused to consent to its dissolution, or even worse, one that refused to accept the king's supremacy. For example, Ramsey was lavishly praised for its loyalty.[102]

If a religious house wanted praise from the commissioners, it had to accept unconditionally the terms of the dissolution. This makes the evidence offered by the commissioners difficult if one wants to find out what really happened in the houses. For us, of course, this mostly speaks about the intentions and politics behind the propagandistic images. The reports tell an interesting story of the world of Henrician propaganda: the king's interest in religious houses was more powerful than their inhabitants' ability to turn his power to their favor.

The reality of the religious life was such that in almost all houses there had, at some point, been some trouble with sexual behavior, and when needed, this could easily be turned against the religious life. Lists of evidence of misbehavior in religious houses are extant from long before the period under consideration. The lists compiled by bishops during their visitations in religious houses, called *crimina comperta*, were often quite long, but they were collected for the purpose of reform from within. For example, in a list from Lichfield there are 25 pages – the most common crimes being sodomy, superstition, and lechery.[103] The crimes mentioned in visitation reports were often insignificant and occurred over a long time span. Though our intention here is not to establish the level of virtue in religious houses other than as images and attitudes, it could nevertheless be mentioned that research on these earlier visitations has shown that religious life was mostly chaste. As an example, one third of the Premonstratensian houses visited at the turn of the fifteenth and the sixteenth centuries had proved impeccable, with no need of reproof. This finding confirms Penelope Johnson's observations that 'better than 95 percent of the cloistered women and men... lived up to their vows of chastity' in the province of France she studied.[104] Calculations by Peter Marshall show that English clergy were no more adulterous than this: only 5 percent of parishes had a priest suspected of fornication at any one time.[105] If people knew this at the time, perhaps the representations read from the texts of the most radical critics do not convey the attitudes of the majority at all.

If the evangelicals expected life to become wonderful and clergy to become sinless after the dissolutions, they were wrong. The conservative policy under Henry, that neither those who had taken a monastic vow of chastity, nor secular clergy, were allowed to marry, was problematic for the evangelical reformers. Some English clerics followed the German example and married, but their action was perceived as illegal. The English certainly understood that if the clergy married, there would be clergy with families, personal clerical property – and clerical inheritance. This was a problem in the Church, which had designed clerical celibacy (among other reasons) to keep the property of the Church to itself. Henry certainly understood this, and did not grant licenses for clerical marriages. It would have required a significant change in mentality to accept married priests, and I am convinced that this did not occur in England during Henry's reign. The supporters of the reformers' cause were relatively few and their influence not yet strong enough, regardless of the noise they made. The counter-argument was no less convincing.

Those religious who wanted to marry caused problems, and naturally, the Catholics were dismayed by their activities in the marriage market. William Barlow suspected these *apostatas* were deceivers of women, and even married several women in different counties. Their 'gorgeous appareyle' and 'swete tales' deceived women into believing that they were wealthy and good matches, but as soon as the marriage was consummated the husbands disappeared leaving the wives desolate.[106] The Lutherans were seen to have created, rather than solved, a problem. Life after the dissolutions was no paradise for the ex-religious and this was widely recognized, sometimes even with regret. Although there were critics who justified the dissolutions because of the sinfulness among religious, there were also those who thought that things were much worse in society without the option of entering a monastery.

Chastity and sexuality were – inevitably and inescapably – central and important issues in religious life, in theology and in monastic thought in general. They were not issues that could be dismissed with a shrug. The sexual nature of human beings caused constant and serious embarrassment in the religious life even more than elsewhere. Many still valued the old concepts of virginity and chastity as divine qualities that would bring eternal joy. Virginity was important especially because of the Virgin Mary, whose role in religious life was incalculably far-reaching in late medieval religion. In a way, the reformers were unarmed against these concepts, and had to transform them for their own use. It is no wonder that chaste marriage for them was a celebrated institution.[107] Chastity was a cornerstone of Christianity; changing this basic concept was never the evangelicals' target. It would have proved impossible and furthermore, it would have left their own religion standing on shaky foundations.

Part IV
Stabilitas

10
Stability and Mobility

Amonge them the sprite continueth we must beleue
that these can not erre / & for no nother cawse
but that they are shauen / and so clothed /
and caried about on mules and charettes /
all thowgh they be neuer so weked neuer so ignorant in
     scripture /
yee though they lacke their comen sense /
and be moore rude them asses of archadie.
                    Richard Brightwell (pseud.), *A pistle*, f. 27–27v

The notion of stability was of great importance in all religious life, even in the sixteenth century. The history of *stabilitas* goes back to early monasticism, when it was felt essential to separate oneself from the rest of the world in order to find God. The choice was irreversible and meant that once the devout had found a suitable location, he or she would never leave it. A monastery would serve as a permanent place of retreat, which offered great comfort to the soul. St Anthony had written that outside a religious house a monk was like a fish on land. Encouraged by him, monks and nuns remained in one place and abhorred the opposite of *stabilitas*, *vagatio*, to which people were drawn by the Devil. St Benedict himself introduced the pejorative term *gyrovagus*, a wandering monk (later, this term would refer to any wandering people, especially itinerant priests). It was believed that people who moved about belonged to the family of Satan.[1] In medieval Christian ethics *vagatio* was a detestable sin, cured only by the virtue of stability. Its negative aspects were further enhanced by the association of vagrancy with beggary and idleness.[2] Interestingly, the *Shepherds' Kalendar* identified vagrancy with covetousness, as if vagrants wandered and begged out of greed only.[3] This seems to indicate that it was thought that stable people had fewer needs and that vagrants were more inclined to resort to all sorts of vices. They were not fully under the control of society, and consequently were suspect and menacing in their marginality.

The original intention of *stabilitas* is well depicted in a story of a monk of Christ Church, Canterbury, who decided to escape from his house. On his way out he stopped at St Dunstan's tomb to ask for the saint's permission to leave, but St Dunstan himself appeared, stopped him and told the monk that he could never leave the house: he would remain and die there.[4]

In everyday terms, stability and mobility were far more complicated issues than St Dunstan's intervention suggests. The stability of any English contemplative religious house was always relative; the everyday business of running a religious house did not easily allow for the total enclosure of its inhabitants. Religious houses had a mission to fulfill, and enclosure without any contact with the outside world would have suited this very badly.[5] In practice, even nuns (whose enclosure was regarded as especially important, because of women's uncontrollable nature) were often seen beyond the monastery walls, since running a convent required regular excursions into the outside world. Moreover, their duties of hospitality obliged religious houses to be open to the world. This worked the other way round as well; no religious house could have survived without the outside world, and, literally, there was no order into which there was no door. Michael Hicks has shown that even the English Minoresses – who were recognized as the most enclosed order for women in England[6] – had a back door used for delivering supplies, etc. The rule ordered the Minoresses to have a door high up on the wall so that it could be reached only by a ladder, but apparently the rule was somewhat relaxed in England.[7]

Part IV discusses the role of (mostly contemplative) religious in the world, and finally their charitable activities. My purpose here is to analyze the aspects in which the religious life was perceived to have relevance outside the monastery gates. Were these activities seen as compromising the religious ideal?

The mendicant orders for men were created to emulate Christ's apostolic mission on earth, and friars were not bound by enclosure. Teaching and preaching demanded being on the move and sometimes even extensive travel. The itinerant mendicants were, however, always members of a certain house and carried licenses to travel. Their movements were thus controlled, so that friars did not become guilty of the sin of *vagatio*.

For a contemplative monk or nun, *stabilitas* had, it seems, by the sixteenth century become synonymous with remaining bound to their monastic vows. Stability no longer kept religious at a certain house, and even transferring to a different religious order was not unheard of. Regardless of the change in definition, enclosure was still regarded by religious themselves as essential and infinitely beneficial; in everyday religious life, monks or nuns were not supposed to be allowed to leave their houses, even for short errands, without a license issued by an abbot or equivalent, and at least in theory, no one was allowed out unaccompanied.[8]

The ideal of stability and especially strict enclosure brought ample and welcome opportunities for excelling in virtue. This is illustrated in a miracle of brother Simon of Chester, told by Henry Bradshaw. Simon was a virtuous monk, who spent his time in contemplation, prayer, and writing. His sanctity aroused envy and resentment among the brothers in the house, and they started victimizing and bullying him. Brother Simon tried to endure this but in the end, his strength ran out. When the behavior of his oppressors became intolerable, he decided to leave Chester and go to another house.[9] The reader who by now may be even too familiar with Chester's great saint, is probably not surprised to hear that St Werburga appears to brother Simon, pacifies and comforts him, and tells him that it is his portion in life to endure:

> sufferaunce shalbe great ioye and pleasure
> And for thy pacience thou maist be sure
> To haue rewarde in blis perpetuall
> At thy departure form this lyfe mortall.[10]

Brother Simon earns his place in Heaven by suffering and staying put in his monastery. Even intolerable stability is better than quitting because the reward in the end will be so great.

In the early sixteenth century, the devotional literature expatiated on the rewards of solemnity. This was something that stability could offer. The monk's cell offered a much better view on the world than great palaces: 'What mayst thou se without thy celle / that thou mayste nat se within?'[11] The cell offered a safe haven, a sanctuary in which to serve God apart from the worldly thoughts and deeds. As such, religious talked about it in terms one would use to describe paradise:

> So it is sure to a Relygyows man to continewe & abyde in his cloystre
> and in his cell ... For ther regnyth peace in the Cell / and without is

awaye of batell & stryf. And therfore as Ierome saith he that desirethe Cryste / let hym seke nothinge ellys in this worlde / but let his Cell be to hym as paradyce fulfylled with swetnes of holy scripture / and that use ofte as for delycis / and reioyce in the stodye of them.... drynke and slepe / but kepe styll thy cell / ledith a monke to his ordre / and so lytyl and lytyll he retournyd agayn to the holy workys of perfeccyon.[12]

Ultimately, in the enclosed world, no one is present but God and the monk himself to control his devotion. Therefore, a cell provides a shelter from evil influences and a haven which sinful thoughts cannot penetrate, if the monk or nun does not personally invite them in. If the monk or nun is tired of temptations and tribulations, only the cell and study there can reinstate him or her in perfection.[13] Religious believed that when a monk left the cell, he immediately exposed himself to the world of Satan.[14]

Donald Logan's study on *apostatas* further shows that in late medieval society, both for the clergy and the laity, the idea of a religious leaving his order was a disturbing thought. Life in the world would lead to perdition, because it was inevitably ruled by sin and the rules of flesh. The fate of the apostate was sealed by official excommunication, which could not be canceled unless the runaway repented and returned to his house to suffer severe penance.[15] The horror of apostasy lived on and re-emerged in some reactions to the dissolutions of religious houses. Prior Thomas of Christ Church, Canterbury was agitated enough to write to Cromwell on the eve of the dissolution that his brothers would never agree to give up their religious habits: for nine hundred years they had served God as monks – they could not return to the world because it offered so many more opportunities for sinning than life in a religious house ever could.[16]

Richard Whitford was bold enough to name the sin that would most likely beset a religious who re-entered the world: the sin of lechery.[17] For him, monks were evangelical eunuchs who had to be kept from all opportunities to sin. The less one had to do with those who remained in the world, the better.[18] Thomas More was of the same opinion. He stressed that it was imperative that monks give up all worldly cares and withdraw from the world so completely that they did not even read letters sent to them from outside. It was their profession to be solitary, contemplate their sins, and refuse to look back at the Sodom they had left behind.[19] As More's words reveal, the monk's and the nun's role was to be contemplative, cloistered, and enclosed. The example of the pious Carthusians certainly must have had an influence on him. Though

More's text argues the case for the traditional ideal of monasticism in a diatribe directed at a heretical monk, he is not merely reiterating something learned: the ideals of cloistering came from somewhere deeper; it was a fundamental part of the way in which he thought and of his culture. For him, the world simply was that way.

Stabilitas seems to be an ideal so rooted in the way people thought of and perceived the world that they did not fundamentally question it. Monks and nuns were felt to be better off enclosed, or at least constrained to a degree. It can even be argued that the evangelicals found the ideal of stability important. They denounced vagabonds and seem to have expected a strict enclosure of the monastics, since they criticized abbots for taking dispensations and leaving their houses for the bishop's miter, taking the sword in the one hand to be used with the crosier in the other.[20] Christopher St German wrote that monks should be 'holly entendynge to fastynges and prayers in the places where they renounced the worlde, and that they forsake not theyr monasteries for no busynes of the churche ne of the worlde.'[21] What else was this but idealization of *stabilitas*?

Stabilitas protected religious from the world and the world from their unwanted interventions. Perhaps it would not be too imprudent to see a trace of arrogance in the words of successful men like St German, men of the world, who could perhaps afford to put priests in their place. More and more, England was becoming a layman's domain, a territory in which old ideals could be adopted to advance a different order. But as part of the mentality of sixteenth-century England, the old religious ideal of stability certainly survived for a long time.

11
Lanterns of Light

The original duty of a monk who honored *stabilitas* was to weep for the world, not to teach.[1] In the course of time, praying, teaching, and preaching became almost equally important aspects of the social side of the monastic and especially the mendicant life. Inevitably, strict cloistering had to give way to these social functions. In this chapter, we will examine in more detail perceptions of the role of the religious in society. The nucleus was the notion that the clergy should be 'lanterns of light' for the people. We will also discuss whether or not religious were believed to be able to live up to their expected role. We will then look at four aspects of monastic life that all were included in the notion of illuminating the world: teaching, preaching, praying and liturgy.

Ideally it was everyone's personal decision in what way he or she chose to serve Christ. Whatever choice was made, it brought with it certain duties. Monks had to pray and teach, and the laity had to take care of worldly business, but also of the Church, because it was the layman's duty to see that those who prayed for him were provided for.[2]

The two lifestyles, the contemplative life of the clergy and the religious, and the active life of the laity, were determined by the character of their relationship to meditation and to the service of God. Contemplative life in religious houses meant continuous service to God, prayer, and meditation. The worldly life of the vast majority was called 'active,' a label inspired by classical authors such as Cicero. This was suitable for those for whom acting in politics or other worldly occupations was a necessity. Ideally, they too could withdraw to a religious house after fulfilling their political and other worldly duties, though this seldom occurred. Active life was not regarded as an easy option, for the challenges it presented were demanding. *Vita activa* was an ideal personally favored by many humanists and reviving yet another ancient ideal fitted well into their program. However, their attitude was not one-sided; Thomas More was not alone in his conviction that monasticism could serve both the soul and humankind. Similarly, early Italian humanists, led by Petrarch, seem to have been relatively enthusiastic about monasticism. Petrarch himself visited the Carthusians of Montreux and wrote extensively on the solitary life; there were others who followed suit.[3] It must be noted that the mass of the people, naturally, could not entertain hopes of a choice of lifestyle. The contemplative life in its monastic form had always been principally a choice for people of relative wealth. In typical characterizations, *vita contemplativa* was defined as complete devotion to God. By no means did this mean passivity, since contemplation was regarded as inward action. It was thought that the eyes of a contemplative person would rise from vain, earthly splendor to the bleeding wounds of Christ. Silence and speaking, fasting, keeping company or staying in solitude were all outward manifestations and only means to achieve the perfect state. The love of God was the most important factor in the contemplative pursuit.[4]

Contemplation was by no means restricted to religious houses, and during the fourteenth century it had become fashionable and even officially encouraged to combine the two lifestyles. This meant that ordinary worldly life included a great deal of contemplation. Active life permeated by contemplative activities was called a 'mixed life.'[5] When people discussed the possibilities of contemplation, they hierarchized three different categories. The first category was love for God without a deeper understanding of spiritual matters. This was considered appropriate for people who lived in the world and led an active life. The second was intended for the educated and for the clergy, since it involved a more thorough knowledge of God and greater study of his word. The third style of contemplation was a complete understanding of God and living in his pure love. This was the most profound of the categories, a form of contemplation that always required complete orientation and concentration on spiritual matters, which in practice limited this kind of contemplation to religious houses.[6] Altogether, this merciful definition of the contemplative life offered everyone the possibility of spiritual excellence and a chance to lead a life of piety.

Expanding the notion of contemplation in the manner discussed above was essential for the survival of the monastic tradition. Nonetheless, it could be asked if the broadest definition of contemplation – as the love of God without a deeper understanding, which could practically be applied to everyone – actually served as a factor in growing secularization,[7] but it is more probable that *vita mixta* was a sign of the development of different forms of devotion, activities in pursuit of a more personally experienced religion.[8] This had nothing to do with secularization. The original idea of *devotio moderna* in the Low Countries was based on a mixed life. Groups of people would live together in contemplation, but they would not form an order or take vows. This form of religiosity seems especially to have attracted women. In England, too, there seems to have been at least one community (in Norwich) for lay women who lived in chastity following the example set by the continental communities.[9] *Devotio moderna* indicates a restructuring of the devotional and religious patterns of the previous centuries. It offered an ordinary person, even a poor one, a new way of religious expression, but it does not allow us to forget the communal nature of all religious practices, since the supporters of *devotio moderna* did form communities. All religious forms and phenomena related to new religious movements, including, for example, literacy in the form of reading aloud, can be seen as strengthening communal devotional practices, recluses being the sole exception to the rule. As Eamon Duffy points out, not even the urban elite isolated themselves.[10] The occurrence of *devotio moderna* coincides with the rise of religious guilds and fraternities at the end of the fourteenth century. Perhaps the trend can be traced back to the thirteenth century, when the laity began to recite the rosary.[11]

The trend toward a shared contemplative experience for all can easily be seen as diminishing the importance of enclosed monastic religion. Traces of the urge to remove barriers between lay and monastic religion can be detected in the humanist Lorenzo Valla's appeal that all devout people, not only religious, should be called *religiosi*.[12] It is interesting that the religious change in itself was not *away* from cloistered type of devotion but rather *toward* it, thereby making it accessible to more people. The movement did nothing to the call for contemplation, and, in the end, it retained old communal ideals. This notion is supported by the history of the Italian women penitents. They too were gradually moving toward more structured communities than their original ideal had been. Instead of living in private homes, as they did when the movement began, these women gradually became members of semi-monastic communities. Interestingly, this process dates back to the fifteenth and sixteenth centuries.[13]

The early sixteenth-century religious movement was in part a product of the climate in which *devotio moderna* was still in many ways valid. For the supporters of the New Learning, contemplation in itself was not an end. Instead, they sought to give new meaning to their relationship with Christ. For William Tyndale, humility in Christ's eyes was crucial. He posited that a person could be divided into two aspects: first, what the person was, and second, what he could transform himself into in Christ's love. Following Christ's example, he could become humble, meek, and patient. A good Christian's goal was to become such a lowly creature that he would allow others to walk over him, be trampled on, and yet love these people. Christ's love was especially important because everything else was wrong in God's eyes and unacceptable in his kingdom.[14]

The clergy's humble duty was to bring a new light of grace to the world. It was the duty of the king to encourage the clergy to shed light on the path where people walked in the dark. This allegory was often repeated in the writings of both evangelicals and orthodox. The ever-skeptical St German asked that if the light of spirituality was darkness, where would the temporality 'then fetche their light'?[15] In his answer to St German, Thomas More made it clear that the clergy was not as evil and dark as some would believe; he even asked how it was possible that there were good men if all those that preached to them were corrupt. How could any evangelical brethren exist? he asks sarcastically:

> that syth this realme hath (as god be thanked in ded yt hath) as good & as faithfull temporalty, & (thought there be a fewe false brethren in a great multitude of trew catholike men) as hath for ye quantite any other countrey cristened, it must nedes, I say folow yt the clergy, though it haue some such false noughtye brethren to, is not in suche sore maner corrupted, as the boke of diuysyon goth about to make men wene / but as good for theyr part as the temporaltie for theyrs.[16]

The criticism against contemporary English religious usually did not include their predecessors who – this seems to have been the belief of even the most outspoken evangelicals – had led a good and commendable life; indeed, monks had brought Christianity to the country. Only the present 'fatte monkes' were guilty of not fulfilling their duties. Even Thomas Starkey conceded limited approval of the contemplative ideal, regardless of his overwhelming preference for the active life. Although he recommended dissolving most of the religious houses, Starkey was willing to let some houses continue for those who were very pious or ill-treated by the world. He described religious houses as safe harbors from the worldly tempests, sanctuaries for prayer, reading of holy scriptures, and meditation.[17] In this sense they could even be centers of light, though his text mostly reveals an attitude of muted admiration.

Many evangelicals did not share Starkey's view. For many, the 'sha-velynges' were far from luminous lanterns; rather they were seen as depravers of spiritual nourishment. Unlike many others, Peter Moone saw the past as utter darkness – a world of sleep and ignorance in which no one had cared for the truth. Not recognizing their blindness, people in the past had trusted in their eyes and been horribly deceived by religious and clergy.[18] This kind of imagery is repeated elsewhere when contemporary religious were discussed. For example, George Ioye was enraged about the handling of the famous case of the prophet and nun Elizabeth Barton. He cursed those monks who had helped her, and accused them of diverting people away from God and toward delusions, lies, and idolatry – all crimes punishable by death.[19] For some, all religious were representations of perversity and corruption, not messengers of hope. John Frith was convinced that Catholics were great persecutors and 'as mercyfull as the woolf is on his praye,'[20] while William Tyndale, for his part, saw the religious life as tyrannical and contrary to religious freedom. As an example, he wrote about the Observants, who forced all their members to live by their strict rule; in Tyndale's mind, the kingdom of heaven should have been ruled by love, not by such oppression.[21]

Many of the supporters of religious were also willing to admit that the relationship of religious to their inevitable worldly activities was not always a healthy one. Erasmus set an example for such criticism by reproaching those religious who put their rule before God's word,[22] but Thomas More distanced himself from such criticism. More was certain that the clergy were able and willing to reform themselves, if there was something that needed reforming.[23] Richard Whitford, again, admitted the existence of problems in the religious life, and provided his readers with explanations and reasons for misdemeanors. He was convinced, for example, that one reason for the failure of the religious vocation in many cases was that parents sent their children and even their handicapped offspring to monasteries too young – some children were committed to a religious life even before they were born.[24]

It was reasonable to ask what the monk's role was, if an adolescent passed as one. Why would parents commit their children to such a life? Perhaps surprisingly, Richard Whitford seems to have thought that early profession not only caused problems, it brought some excellent benefits as well. He stressed that very young children learned more easily than older ones and could easily be 'broken in good maners: than in forther age.' Like a vessel that would forever smell of the substance that first filled it, a person would always remember what he learned as a child, and in religious life the child could get the most perfect education.[25] Whitford thought that if there was even a trace of repentance, a somewhat fallible religious life was far better than the dreaded apostasy. In this issue, it seems, devout Catholics were adamant: if someone suggested that religious should discard their religious habit, he or she was branded a breaker of holy vows, an enemy of chastity, a wrongdoer, and seducer from the faith, especially by their religious brothers.[26]

It is now time to move on and examine the various roles of religious toward the world. The first glimpse of the different duties of religious will come from education. The Dominicans were renowned for their educational activities, but long before the emergence of the friars, some religious orders, led by the Benedictines and the Augustinian Canons, had maintained large networks of schools.[27] Many children of the nobility were educated at home.[28] For girls, nunneries often offered the only chance of a more formal education, but these institutions were largely restricted to the relatively wealthy, though most houses would take some pupils on charity. (It should be noted that boys could also be taught at nunneries.) The monasteries were naturally paid for their services, and in return, the children were taught to read and write.[29] In theory, monasteries could take students destined to be professed as monks and nuns only, but in practice, teaching lay children seems to have been widespread. Teaching in religious houses continued until the dissolutions, the only significant change taking place in the intake of the pupils: the children they took came increasingly from the gentry, and more seldom from the nobility.[30]

Eileen Power's classic study of nuns concludes that the English nunneries were not deeply educational in their interests. She did not find nuns very literate either. Their libraries usually were quite modest, including essential devotional texts only. Furthermore, nuns did not usually have a knowledge of Latin.[31] Maurice Keen argues, however, that although the education of nuns was not a priority for the Church, nuns could usually read English. There were also acknowledged exceptions to the rule: the Brigittine double-monastery of Syon, for example, valued education highly and the nuns there were better educated than it has been assumed.[32] Penelope Johnson believes that the education of nuns was not intolerably weak.[33] As evidence, we have some examples of continental nuns, such as Caritas Pirckheimer, known for their excellence in learning even in the age of the Reformation.[34]

It seems that sixteenth-century English gentlefolk respected the education offered by the monasteries, and for many it was a natural choice for their children. Among them is the notable example of the Lisle family whose correspondence in the 1530s reveals a positive attitude to religious houses.[35] Education was certainly regarded as important, for an ignorant

person was like an unreasoning beast.[36] Everyone had to have at least a
sense of the order of things, and in the early sixteenth century, the clergy
were commanded to teach everyone the articles of the faith at least four
times a year. This was especially important to those who did not speak
other languages than English, because they could not study the Bible on
their own.[37] It is possible, as Alister McGrath suggests, that the importance of education was newly recognized in the fifteenth century because
of the influence of such religious movements as *devotio moderna*.[38] The
humanist ideals of education should not be overlooked either.

Educational interests should not be seen as manifested only as concrete teaching. There were many other aspects of the religious life that
could be included. For example, religious plays served as an important
educational medium, as did writing and copying devotional texts. As
Michael Sargent has shown, even the most enclosed of monastics, such
as the Carthusians, fulfilled educational functions by copying the Bible
and liturgical works, writing new devotional texts, and distributing
books. On the eve of the dissolutions, regardless of the fact that the
printing press was already well established, people were not yet able to
see the profound changes the press was to bring to literary culture, and
thus the importance of monasteries as literary centers had not yet
significantly diminished.[39] Lastly, it must be noted that monasteries
not only produced literature, they also preserved it in their libraries.[40]

The educational activites of the religious orders came in for mockery
and criticism from the evangelicals. For the most critical, the 'monsters'
of monasteries prepared children for a life of evil, idleness, and gluttony,
instead of preparing them for an honest profession which would profit
the common good, as lay upbringing did.[41] There were, however, also
reforming voices that stressed the importance of religious houses as institutions of education: the justification for their existence would then lie in
their very teaching, since monks had to have an active role in society.[42]

Following the medieval tradition, many presented, however, the
monks as ignorant and thus as good targets for ridicule.[43] Erasmus
stressed their widespread ignorance and insinuated that monks understood nothing. This was the reason why religious life was so cold and
flaccid:

> Neyther I thynke, any other thyng to be the cause, why we se ye
> charitable liuing of our monkes and cloysteres so to fayle euery where
> / to be so colde / so slacke / so faynte and so to vanysshe away, than
> that they contynue all theyr lyfe, and wexe olde in the lettre: and
> neuer enforce to comme to the spiritual knowlege of scriptur.[44]

Religious' knowledge of foreign languages was also criticized. Friars' stumbling Latin was the butt of humour.[45] A good example of this comes from a piece of advice to anxious husbands. Men are wary about teaching women Latin and Greek, because they are afraid of their wives becoming too conversant with the clergy. The author comforts the husbands and reassures them that teaching the women will only serve the husbands' own best interests, since the ignorant clergy will have absolutely nothing to do with a woman who is cleverer than they are themselves. Becoming more learned than the clergy is not difficult, they scorn. Consequently, learning will actually keep unwanted clergy away from wives:

> if their wyues coulde Latyn or greke than myght they talke more boldely with preeste and freres / as who sayth / there were no better meanes (if they were yll dysposed) to execute their purposes / than by spekynge Latyn or greke / outher else / that preestes were commenly so well lerned / that they can make their bargeyne in latyn & greke so redily / whiche thing is also farre contrary / that I suppose nowe a dayes a man coulde nat deuyse a better waye to kepe his wyfe safe from them / than if he teche her the latyn and greke tonge.[46]

This is a perfect trick: the adulterous priests and friars will avoid learned women, the women's honor will remain intact, and the husband will escape cuckoldry. This was, of course, only jesting and cannot be taken at face value as evidence of clerical learning. Priests were expected to master Latin to some degree at least, and failing to do so brought deserved criticism, but not everyone was that ignorant.

Similarly, the language skills of nuns were so bad, in the evangelicals' opinion, that nuns were exhorted to use their time more profitably than trying to mumble a language of which they knew nothing. They should have read a language they could understand, rather than spend hours singing in Latin:

> No maner of spirytuall ioye can theyr herte manye tymes, serchynge nothynge ells but vayne glory. It were moche better for them to rede theyr houres in a language that they vnderstode.[47]

Alongside teaching, preaching was one of the main duties of some religious. The Dominicans were specifically known as the preaching order (*ordo predicatorum*), and many ordinary people favored the preaching orders over the strictly enclosed ones precisely because preaching brought them closer to the people. They also suggested that those monks and friars who were licensed to preach were likely to know the Gospel better than others.[48]

Because friars were freely granted licenses to preach, they traveled the country and were seen and heard frequently. They preferred cities simply because they offered more people to be instructed than the countryside. The evangelicals often acknowledged the skills of the preaching friars, which is hardly surprising if we remember that many of the evangelicals were ex-friars themselves. It is interesting that in Germany about 40 percent of evangelical preachers were former members of religious orders. Scribner suggests that part of Luther's popularity stemmed from his public persona as a friar bringing the Bible to the people, and it seems likely that mendicant preaching styles could account for the effectiveness of many a popular evangelical preacher.[49]

In the light of this, I would like to suggest that there was still a mentality which acknowledged the holiness of religious; even though the interests of friars were sometimes transferred to the new learning, they were still mendicants in their skills and techniques, and their targets remained the same: winning people's souls.

That said, it should be recognized that English evangelicals voiced stong criticisms regarding preaching by religious. Simon Fish thought that monks were totally uninterested in preaching. Because he seems to have looked at all questions from an economic viewpoint, he took wealthy Westminster Abbey as his example, and claimed that people had paid for so many sermons and masses to that house that not even a thousand monks would have been able to carry them out. They happily accepted the money, but did not care at all whether the donor received anything for the money spent.[50] Friars, on the other hand, could be attacked for preaching for money only. Their activities were dependent on their pay, so if friars were successful in their begging, there would be sermons, but alas – if not:

> There wolde not be in seuen yeres,
> A sermon in the poore contry.[51]

Nicholas, a Lutheran-minded character in the dialogue of William Barlow, offers another criticism, stating that friars run around the country preaching old-fashioned stories that profit no one:

> to beginne furst of all at the freers / ye se what a rascall rable ronneth abowte the cuntrey wyth bosomed sermons, preachyng fables and

olde wyues tales in stede of the worde of god...to the slaunder of crysten relygyon.[52]

Nicholas criticizes the subject matter of the friars' sermons, seeing them as evidence of old superstitions and beliefs. For some, rather than the subject matter it is the techniques the friars use, such as their shrill preaching voice, that deserve ridicule.[53] Sometimes it was the length of the sermons that was laughed at, but this criticism certainly applied to parish priests as well.

In his *Debellation* Thomas More repeats an amusing *exemplum* of an impudent woman and a preaching friar. The woman, wearing a red hood, keeps whispering to her pew fellow while the friar is preaching. The friar catches her at it and asks if the woman in the red hood would stop her 'babbling.' The woman, infuriated by the friar's insolence, shouts back that it is really the friar who babbled the most: had he not done so for the past hour?[54] The last laugh in More's story, however, is not on the arrogant friar, but the red-hooded woman, who makes herself ridiculous by behaving badly.

Traditionally, the prayers of monastics had been regarded as exceptionally powerful, because monks had a special relationship to God. It is clear that in the long run the evangelicals' denial of purgatory removed the need for monastic prayers for the soul; since the path of personal devotion led directly to the kingdom of heaven, a praying order was no longer relevant as an intermediary. In the 1520s and 1530s, the question of purgatory, especially the problem of justification by faith, caused polemic disagreement in England.[55] Thomas More especially reacted strongly to the critical writings, and it is clear that for him among the worst of the new heresies was the skeptical attitude toward the existence of purgatory. He could not forgive the evangelicals for challenging these articles of faith. More believed, like many of his contemporaries,[56] that the prayers of the clergy and religious did help people to move out of purgatory more rapidly.

More's message to non-believers was clear: the road of heresy was a direct path to hell. To spite the evangelicals, More exhorted his readers to prepare for their death in good time, and put away money for masses for their souls to ensure that death would not catch them unaware. The medieval ideal of a peaceful and well-prepared death, in which religious played a crucial role, was clearly what More had in mind. He appealed to the conscience not to let anyone die without ensuring the care of his or her life beyond: paying for masses was an act of love. More argued forcefully that anyone who left his neighbor to burn in agony in purgatory, when he could help simply by offering suitable prayers and masses, was a brute. Could anyone enjoy a good dinner while watching someone else burn in agony? he asked.[57] More's words convey that for him this was not merely a religion of old habits, but a living faith.

Some years later Thomas Starkey argued in a letter to the king that after the first dissolutions there was genuine anxiety among the people when they realized that the prayers for their dead were now lost. Starkey clearly avoids stating that the monasteries had not carried out their duties as mediators; rather, he tries to convince the king that it was much better both for the living and for the dead that everyone, not just a handful of monks, should pray for the dead in 'charitabyl memory.' The position he was putting to Henry was that the dissolutions would not harm anyone or contravene the last will of the deceased, since it was much better that a multitude instead of a few religious should bear the responsibility for dead souls. He was certain that if the dead rose to see the present state of decay, they would be more than willing to turn their religious foundations to more beneficial use. The houses would then provide an 'ornament to the commyn wele & not as they be now sklaunderouse & therwyth grete detryment.'[58] So pressing is Starkey in his reasoning that it begs the question: did Henry perhaps have doubts about the religious outcome of the dissolutions? Suppressing religious houses, even though it could be justified, was a matter that a king known for his religious nature and for the pious history of his family would consider carefully. Perhaps it is in this context that Starkey does his best to reassure him and everyone else that no one needs to worry:

> by thys acte of suppressyon they [the dead] suffur [no] wrong or iniurye, but rather, as fer as mannys reason may attayne, schal take grete consolatyon to see theyr possessyonys, wych long haue byn abusyd to the nuryschyng of vyce & idulnes, now conuertyd & turnyd by your gracyouse gudnes & wysedome to the commyn comfort of theyr posteryte & to the settyng forth of goddys glory, the wych dowteles ys more schowyd & openyd to the world by the multytude & increse of hys pepul lyuyng togyddur in chrystyan cyuylyte then by a few lyuyng in the monastycal lyfe & solytary.[59]

Starkey's last words are significant. The multitude is better than a way of life that seeks isolation, the world is better than a monastery: God's glory is better shown and his will better fulfilled by multitudes than by individuals. As the multitude is committed to serving the multitude, they will never concentrate on personal achievement, as individuals inevitably do, Starkey seems to reason.

It should be noted here that although the evangelical current was turning away from the religious communities, some hard evidence shows that ordinary people remained loyal to them. The friars, especially, had gained a strong position in the people's hearts.[60] Peter Marshall sums up the evidence of requests for masses in wills and comes up with an interesting figure of 45 of the testators (of 1,707 wills) requiring masses specifically from regular clergy, of which the majority, 37, favored friars.[61]

The friars often won the contest for souls, in particular, over the choice of confessor. The relationship between confessor and penitent could become very close, and the confessor could have great influence on the people he guided spiritually.[62] This is why (discounting financial motives) priests competed so keenly with friars for souls. These relationships are too important to be overlooked. It was sometimes alleged that friars were popular confessors because they gave lighter penances than others, as a form of competition. This can be illustrated by means of an *exemplum* of a mother who cannot tell her son that he will be punished for being late to school after oversleeping. The mother comforts the frightened child and promises him that he will not be beaten at school; all he needs to do, she says, is to take his bread and butter and go. She lies to stop the boy from crying, because she cannot bear to see his agony. But, of course, the boy is beaten for being late and suffers all the more. Such false assurance was felt to be harmful; compared to the pains of hell, hurt feelings were hardly serious.[63] Easy penance was not enough in God's eyes and could harm the penitent's soul, and at worst send him perdition.

The fourth role of religious requiring consideration here is their ceremonial duties. In the early sixteenth century, the term ceremony was defined as 'a rite, custome, usage and an outwarde form or behavior' in the service of God or in his honor.[64] Doubts were expressed about the ceremonies of the Church; some thought that religious howled at their services like hunters:

[they] syng in theyr querys wyth as great deuocion, as hunters do at the halowyng of a foxe / hauyng delite in yellyng of theyr voyces and cryeng of organs, but no swetnes of spyrytuall melody.[65]

Likewise, matins, prime and the other hours were roared and mumbled so laboriously that prayers were altogether forgotten.[66] Those who were dissatisfied with this ritualistic worship felt that it was all designed for the ceremonies' sake, and not in God's honor. The ceremonies were felt to have become habitual, a routine that lacked the appropriate devotion and fervor, spirit and free will; only superstition, pride, and hypocrisy remained:

> Of their fantastycall ceremonyes.
> God and his lawe they omytte,
> Applyenge their malycyous witte,
>   To kepe mans inuencions:
> They are patrons of ydolatry
> Promooters vnto heresy,
>   And bryngers vp of dessencyons[67]

In response to those who felt this way, Richard Whitford argued that ceremonies could never be harmful if they were not excessive.[68]

The counter-argument to this was to argue that many of the rituals were no more than superstitions. Services had become too elaborate.[69]

For many, criticism of the forms and formalities was a sore point; a ritual belief, as it were, was built up with images, vestments, and gold chalices. Form and ritual were positively perceived in the religion of ordinary people. Henry Bradshaw mused upon the fact that his beloved St Werburga was richly honored and remembered with rich men's donations in vestments, gold crosses, books, and bells, and by poor men's candles and towels. In return, to show her gratitude, she performed miracles and helped people, especially women and children who were close to her heart.[70] Veneration was crucial to the traditional faith, and any attacks against images, for example, delivered a severe blow to its very foundations. The evangelicals felt that ceremonies had taken the place of virtue and holiness, that the spirit had been replaced by the flesh, and that reading the hours, singing, kneeling, fasting, celebrating masses, and other ceremonies, had replaced humility, contentment, sobriety, patience, and poverty.[71] From their point of view, religious knew nothing of a spiritual heart, since they had never had a taste of sobriety 'for they abyde alwayes in the flesshe and in the letter of theyr rule & of the commaundementes.'[72] The evangelicals detested religious for concentrating on the surface and for neglecting the deeper aspects of devotion.[73]

Although some critics found the ceremonies amusing, as the priests went around the altars 'as it were a certeyne of trees,' made offerings, sang, clattered bells, and ran 'aboute the chyrche' until they sweated and became hoarse,[74] these manifestations were crucial for ordinary people, whose daily worship was centered on ritual. There is written evidence of these feelings; among them the York Articles produced by the Pilgrims of Grace during the uprisings in the North. They should be read as manifestations of approval of the old religion, and further, as declarations of people's need for their customary ceremonies. They were felt to be essential in all aspects of life, and indeed for the survival of society. The declaration of the rebels stated that without monasteries everything would change dramatically, and not for the better.[75]

The question of justification by faith was one that inevitably generated discussion on the necessity of ceremonies.[76] If only inner faith (and God's mercy) was needed to enter Heaven, ceremonies became vain and useless, and the old faith perverted. The symbols of the pope's religion almost automatically shifted into symbols of evil and darkness:

> Here of commeth it, that we haue now the abhominacion, of the popes power, of pardons, of masses for the deed and quycke, of merites, power and intercession of sayntes in heauen, of worshyppinge their bones vpon erth, of ydols, and vayne ornamentes, pompe and prayenge in the temple, and of the whole swarme of ydele religious.[77]

The whole flock of idle religious would be made useless at a stroke. This is what happened when the religious houses were suppressed, of course, but almost certainly not for these reasons. Henry's motivation was much more mundane.

Everyone, evangelicals and conservatives alike, seems to have agreed that it was the responsibility of parishioners to look for decent clergy, to consider carefully what kind of clergy they chose as their spiritual guides, and what kind of ceremonies or discipline their religious houses actually kept up.[78] Thomas More, whom we have mostly seen defending the religious life, also criticized the religious orders for preferring their own ceremonies to the will of God.[79] The separation was sometimes taken so far that different orders formed sects, as it were, each with its own specific ceremonies.[80] The 'sectarians' were, as Thomas Solme puts it, 'the vpholsters of Antichrist the son of perdicion' – they were satanic excrement, the very 'dropingis of the deuylis tayll.'[81] These sectarians would gladly deceive people with their ceremonies and fake miracles, he said.[82]

As we have seen, differing visions of the roles of the 'lanterns of light' raised many questions during the first half of the sixteenth century: was their role to withdraw from the world, or to what extent should they participate in its bustle? When this became a question about the old faith and a reform of that faith, an inevitable collision followed. The evangelicals often found the practices and practitioners of the old faith in need of reproach, and the reverse was certainly true. For someone like Thomas More, evangelicals were a 'barrell of poysen' which the devil had 'late sette abroche, and laboreth by them to poysen other men,' and their heresies 'hepe vp to the skye theyr foull fylthy dunghyll of all olde and newe false stynckyng heresies.'[83] Evangelicals were not very numerous in England at this point, but some of them were vociferous. This is corroborated by More, who insisted that the realm was not full of heretics – they were simply noisy: 'for all theyr besy bragyng, they be yet in dede but a few.' For More heretics and people such as Christopher St German[84] turned people against the Church, incited riots, set the laity against the clergy, armed the people against magistrates, and conspired to raise rebellion against princes. More's belief was that there was no real grudge between the clergy and the laity. The evangelicals were too eager to criticize the clergy and forget about their own sins. More notes that

> so charming does everyone find his own notions, so fragrant does everyone find his own farts, that while we wrinkle our foreheads at other men's jokes and condemn them as bitter we cherish our own even when they are not as amusing and more bitter still.[85]

More seems to have been down to earth regarding the question of the human nature. He thought that one bad priest did not mean that everyone was evil and sinful, and he regretted that people were so eager to hear damning stories about the clergy; such news would 'fede and nuryshe any suche delyte.'[86]

Having started this chapter with a discussion of light and darkness, it is perhaps suitable to conclude with the same theme, and remain in the company of More a while longer. He advised people, when they wondered if the clergy shed enough light, to turn to the words of St Chrysostom and Christ himself. The saint had pointed out that if priests were corrupt, all virtue would be swept away from the world, and of course Christ himself had asked how dark would darkness be if light were dark. More concluded that if the clergy were evil, the laity must be even worse.[87] It seemed self-evident to him that this was not in fact the case.

More was not the only one to hold on to his old beliefs, rituals, and prayers – people did so for years after the death of Henry VIII.[88] Organized religion proved difficult to reform even after (or perhaps because of) significant burnings of books and images.[89] Someone, after all, still had to teach, preach, and pray, and someone had to perform rites to keep society going, and life purposeful and peaceful. Even though it is a cliché, sixteenth-century people's lives were still permeated by religion and the Church, to the extent that radical and sudden changes in religious practices were very difficult. It must be remembered that the dissolutions, in the context discussed here, changed relatively little. There were ex-monks and ex-friars, cartloads of qualified people available for all the duties of religion: for carrying out sacraments and religious rituals, for preaching at village churches, chapels, and chantries, and all this mostly in the old style. If people confessed to an ex-monk, all the better. One could never be certain – perhaps they still carried that special aura of religiosity and the mystical qualities about them that could bless anyone in their vicinity.

12
The Duties of Christian Love – Charity in Action

Charity, God's love, was understood to be separate from earthly love to which sin and sinful feelings were inextricably linked. This division probably came to English from the classical languages: in Greek, the word *agape* meant charity, and *eros* love, and in Latin, correspondingly *charitas* and *amor*.[1] Charity was the true key to heaven: 'of all vertues charyte is moost vertuous & moost necessary to mannes soule';[2] without it, preaching and teaching, prayers and alms, pilgrimages and penance were all useless. An *exemplum* of an Irishman provides a warning: this man gives such a large portion of his riches to the poor that everyone thinks of him as a saint, but after his death he appears to an acquaintance blackened and stinking horribly – regardless of all his famous charitable activities, the rich man has been cast into hell, because everything he had done had been for vainglory, not for the glory of God. God, however, cannot be deceived.[3]

Even though the term charity did not have as restricted a meaning as it mostly does nowadays, a major part of charity, even in the sixteenth century, was concentrated on what sounds familiar to us: doing good deeds, helping the poor, etc. Needless to say, charitable deeds were expected not only of religious; they were everyone's responsibility. However, much organized charity was channeled through religious institutions, abbeys, convents, and hospitals. In late medieval society helping one's neighbor in need was a Christian obligation that formed a firm, supporting construction around society. Judith Bennett and Giovanni Levi have investigated the ways in which these constructions in practice could save people from disaster.[4] Because lay charity usually did not extend very far from the family circle or the closest social referents, however, various religious organizations had to take responsibility for those who already were on the margins. Inevitably, money spent on charitable activities formed a considerable part of the annual expenses of most religious houses,[5] which was just as well, since as individuals all religious were to have deep roots of meekness, water them well, and extend the branches of charity, for he that 'hath charyte hath all goodnesse, & maketh all others goodnes his goodnes.'[6] Charity and piety were inextricably linked.[7]

Were religious represented in this manner by their contemporaries? It seems that once again the past had a rosy hue. Many felt that charity had been perfect in the past: justice, mercy, true love, and amity had been abundant, and dissimulation, pride, and envy unknown. Bradshaw wrote that charity had been fervent, increasing day and night.[8] The past was seen – if the anachronistic term can be excused here – almost in romantic terms. The poor had needed the help of religious houses, hospitals, and almshouses, and had received it. It was recognized by many that the largest monasteries offered systematic charity. In Westminster Abbey, for instance, an almoner was appointed to receive the petitions of those who needed alms and then distribute them. He distributed the leftovers from the refectory daily, and distributed clothing at Christmas. It was the subalmoner's job to visit the sick of the surrounding area to comfort and help them.[9] Along with religious guilds, hospitals, prisons, and parishes, the religious institutions were important for people who wanted to donate money to the poor: they acted as a distributor of food, goods, and money.

Nursing the sick was just as important as giving alms. The first imperative for religious was to take care of old or sick monks and nuns, but religious houses were in fact centers of medicine and had a more varied role in nursing.[10] Many of the largest hospitals in England were directly under monastic rule. A good example is the famous St Bartholomew's Hospital in London's Smithfield, which operated under the Priory of the same name. It specialized in taking care of pregnant women, housing orphans until they were seven years old, and treating epileptics, eye and ear patients, and those who suffered from strokes, edema, fever, or mental illnesses. Many religious houses would also take care of those who were mentally or physically handicapped.[11] According to William Gregory, the London religious houses and hospitals were understood to have played a vital role in relieving the suffering of the poor and other needy of the area: the blind, lame, and the invalids[12] were offered masses, clothing, barbering, laundry, and food and drink. In Gregory's opinion, their activities were charitable and discreet.[13] William Copland's *Hye way to the spytell house* seems to confirm this. He represents a hospital as a center of bustling activity in which the poor were fed, the sick helped, the injured given relief. When the life of mortal toil came to an end, the hospital would even provide a decent burial for the poor:

> Forsoth they that be at suche myschefe
> That for theyr lyuyng can do no labour
> And haue no frendes to do them socour
> As old people / seke / and impotent
> Poore women in chyldbed haue here easement
> Weyke men sore wounded by great vyolence
> And sore men eaten with pockes and pestylence
> And honest folke fallen in great pouerte
> By myschaunce or other infirmyte
> Way faryng men, and maymed souldyours
> Haue theyr relyef in this poore hous of ours
> And all other which we seme good and playne
> Haue here lodgyng for a nyght or twayne
> Bedred folke / and suche as can not craue
> In these places / moost relyef they haue
> And yf they hap / within our place to dye
> Than are they buryed / well and honestly.[14]

This hospital helps those who cannot work: invalids, women in child-bed, those who are injured by violence or suffered from diseases, travelers, and maimed soldiers. Copland also shows that these institutions can face problems caused by many who want to take advantage of them without true need or urgency: some fake being lepers and even put soap in their mouths in order to look very ill.[15] If the charity of the religious houses was worth deceiving for, it must, at least in sixteenth-century terms, have been thought of as helping people. It must have functioned efficiently, and it is probably only for propagandistic purposes that the anonymous author of *Rede me frynde and be not wrothe* writes:

> Of their brothers vexacion,
> They haue no compassion
> Despysynge those that be in sycknes:
> Agaynst all order of charyte,
> They dysdayne for to haue pyte,
>   Vpon them that are in destres.
> To shewe all their vnhappynes,

> So abhomynable and shameles,
> It were over tedyous and longe[16]

Regardless of the order, all religious houses were under the obligation of hospitality, which in practice meant that they had to admit visitors and travelers who came to the gates. Special guesthouses were built for them: they were given food, drink, and shelter for the night. This had to apply to everyone. The abbot's duty was to see that the poor did not suffer while rich visitors were well taken care of. They also had to ensure that people who left the monastery gates had food with them.[17] Copland notes, however, that not all did their share in helping other Christians; there were priests and clerks who preferred to drink and ignore their duties

> at the alehous for to syt bybyng
> Neglectyng the obedyence to the dew.[18]

But apparently most did their duty.

The question of charity was a topic of a lively debate. The issue of justification by faith, being one of the major questions of the reformers, came alive in the form of charity. Deeds of mercy had been a cornerstone of Christianity, as we have seen earlier, and it was difficult to determine a middle way. William Tyndale wrote about justification by faith only, while his friends were demanding more good works to be done for the needy.[19] Some people felt that religious houses could have done more to help the poor. One critic stated that monks should spend the energy used for their multiple services and ceremonies to 'ye humylyte, benevolence, and charite toward their neighbour,'[20] but three or four cases in which the religious were criticized *for fulfilling* their obligation for charity can also be found. Let us take one example. Jasper Fyloll, Thomas Cromwell's commissioner, who was keeping a close eye on the London Carthusians during the troubled times there, reported them wasting fish, bread, and ale, and even distributing food to strangers 'in the butterye and at the butterye dore...which can not be.'[21] Fyloll probably did not want to see this as a charitable action, and so it ended up as a part of his negative report to Cromwell. In the rage that Fyloll certainly felt against the obstinate Carthusians, who rejected everything that was suggested to them, the monks must have appeared simply unreasonable to him. It must have been easy to forget the religious foundations of their inaction and action, not to speak of the propaganda value of their actions if properly used against them.

This kind of accusation were rare, however. Usually those who wanted to slander religious accused them of carrying the best food from their houses to their harlots and relatives,[22] or giving better shelter to wealthy travelers than to beggars:

> Hospytall abbeyes thou fyndest but fewe
> Although some of them for ashewe,
>   To blyndfelde the peoples syght:
> Parauenture wyll not denaye,
> Yf a gentle man come that waye,
>   To geye hym lodgyng for a nyght.
> But yf poore men thyther resorte
> They shall haue full lytell comforte,
>   Neyther meate, drynke, nor lodgynge:
> Sauynge other whyles perhappes,
> They gette a fewe broken scrappes
>   Of these cormorauntes leauynge.[23]

As we can see, religious are accused of deceiving people. It is said that they and their rich guests are like cormorants who barely leave a few bones of their festive meals for the poor men to eat. If we remember what happened to Robert the Devil, it can be repeated here that eating leftovers was something dogs did, and was degrading to humans.

Of all the criticism against religious concerning the question of charity, the friars seem to have received the lion's share. Evidently, they sometimes caused problems since they wandered and begged for their living and demanded upkeep during their travels. Because they had to beg themselves, they could not help the poor in the way or to the extent that the contemplative houses did. We discussed some of these aspects in chapter 4, but we need to briefly return to them here. There were voices that claimed that friars lied when they said that they lived only by begging,[24] and that in reality they accepted anything but land. The fact that they would not own land made no difference, because the religious already had more land than the whole nobility of England.[25] Thomas More found preposterous the accusation that friars received thousands of pounds in alms annually. He claimed that if the religious houses gave all their possessions to the poor, they would certainly be criticized for that as well. The critics would then say that alms were given for hypocritical reasons, and in the end someone would certainly note that everything that would help good yeomen was given to 'noughty beggers.'[26] More seems to have felt that people were never satisfied with anything they had or received, and that the complaints should not be taken too seriously because of this.

The accused reminded their critics that if there were no beggars to be seen in Germany it did not mean that all Germans were wealthy. William Barlow explained that the reason for their nonexistence was the lack of charitable institutions which could help them in their misery, and the poor were thus forced into working for wages that could not support them. The poor in Germany had to be used to a 'sober dyet, and longe exercysed with laboure.'[27] Barlow felt that beggary was better than near slavery with wages that could not buy food: at least it was easier on the body. His logic was based on the need to keep the religious houses alive and wealthy. The reformation in Germany thus sounded a warning to all Englishmen that if the monasteries and hospitals were abolished and not respected, social catastrophe would follow. The confiscation of the property of the German religious houses had not benefited the poor; even though promises were made to use the funds for charitable purposes, the money disappeared into other hands.[28]

The Lutherans were confident that the activities run by religious houses could be transferred to other institutions. Perhaps there would be no begging whatsoever if there were no idle monks and friars to consume the food that should have belonged to the poor. The evidence shows, however, that people were aware of the charitable activities of the religious houses and that they were for the most part appreciated. The plans for using the property of the Church for these ends was well founded and many believed that it would help, but the German undercurrent seems to have shown that improvements were just a dream. More conservative people were convinced that the old system was good enough, and the reforming minority was eager to try something new.

Charity was probably the most concrete activity in people's dealings with religious. The poor would get help from them, and even the wealthy needed their services when they were travelling. It was not a facet of the religious life that could simply be shrugged off. The reformers would have liked to see other charitable institutions in the place of the monastic properties, but such improvements were slow to follow, which caused them great disappointment. After the dissolutions, many longed for the monastic days and the monastic poor relief; it was long into the Elizabeth's reign before the issue began to settle.[29]

To conclude: charity was the most powerful and the most significant virtue. Without charity it was pointless to try to reach God, because only love transported one to holy bliss. Religious houses were understood to be charity in an institutionalized form; they took a load off the shoulders of ordinary working and farming people. This was why religious were indispensable for the majority: without monasteries, convents, poorhouses, and hospitals, life would have been less secure. When they were suppressed, they were sorely missed.

We have now examined several different roles that religious had in the world, and seen that it was not a straightforward matter to find replacements for the services they offered in a world without religious. It was generally accepted that the roles religious could have – praying, teaching, charitable activities, enacting ceremonies – were approved of as good and beneficial. Because they were fundamental virtues and deeds, it was not easy to disord them. The evangelical brethren therefore were forced to concentrate on wrongdoings once again.

The role of religious in the world was a dual one: they had to renounce it only to shortly reappear in it in the form of a teacher, priest, confessor, doctor, or almoner. Religious had roles that were woven deeply into the structures of society and the reformers were unable to solve the problems arising from the hypothetical question: 'what if religious were no more?' There was no significant paradox between the ideal of *stabilitas* and the activities outside religious houses. The poor had to be helped, people had to be taught, and they had to be given spiritual consolation. After the dissolutions this was all too apparent: the new order could not work in an instant, and not everybody could cheer at the disappearance of the charitable and other important activities in which the religious people were deeply involved.

Conclusions

The more helth he hath / the more he complayneth
The more hardy he is / the more he feyneth
The more he loueth / the more he payneth
The more he is beleuyd / the more he lyeth
The more he hath wherwith / the lesse he contenteth
The more he is reproued / the more he murmureth
The more hye of pryce / the lesse tyme abydeth
The more mony he hath / the lesse hym sufyseth
The more vnderstonynge / the lesse he well sheweth
The more he hath done amisse / the lesse he feryth
The more he contynueth the worsse he lyueth

Somer and wynter, sig. A3v

Discontent and satisfaction are human driving forces, opposites that create similar expectations. Sixteenth-century people knew that they should be satisfied with God's will, for it was considered a waste of time to pursue anything other than that. But what was God's will? The answer to this question was not unambiguously clear. It was agreed, however, that eating and drinking, giving and taking, begetting children, owning things and earning money, being respected and honored were actions acceptable to God. It was likewise acknowledged that God's word did not allow gluttony, lechery, enjoying the pleasures of the body indecently, worldly glory, or a proud wish to become mighty.[1] At a very basic level, everyone had similar rights and similar duties toward God. Medieval Christianity had typically regarded the monastic pursuit as the supremely commendable way of life, since in monasteries one could concentrate solely on serving God.

149

Some felt, however, that virtues such as obedience had been wrongfully monopolized by the monastic order, and that their excellence was unduly emphasized; they argued that the path to God in the fields could not be worse than that in the cloisters. Some went so far as to claim that the lay life could even be the more meritorious one. The evangelicals emphasized – and it is noteworthy that in this point they did not go against the traditional learning of the Church – that one could achieve anything God wanted, without becoming a monk or a nun. A simple, ordinary life was meritorious in the eyes of God, though everyone acknowledged that a good life in the world was by no means easy. Just as a monk had to think first of his community and only secondly of his own interest, life in the world required that one put the common good before personal interest. This, for a sinful soul, never came automatically.

The dialogue between the evangelicals and their opponents often became a futile tug-of-war, and at times it is very difficult to uncover the core of the attitudes and ideals beneath all the rhetoric. In what could be called a contest for souls, both the evangelicals and the religious conservatives were equally guilty of contempt for their opponents. Both groups were also capable of making compromises. For example, not all evangelicals condemned the monastic pursuit, but because they saw life in the monasteries as mostly drinking and banqueting, they made it clear that religious were wrong in assuming that the mere superficial expression of religion made them holy.

Most people, in fact, were willing to admit that even though they regarded the religious life as the ideal, it had to be accepted with reservations. A good lay life was far better than a failed religious life, but many would still agree with Thomas More who wrote: 'the same soil produces both wholesome and noxious herbs.'[2] Could it be that those who denigrated the religious life were actually envious of the role of religious and of their relationship to God? Could St German's notion of the ways in which religious regarded themselves as better than others be interpreted as a kind of spiritual envy? It could be argued that many showed signs of envy when they blamed religious for pride in their excellence.

Such envy could well have been triggered by the likes of Richard Whitford, who could not stop expressing his delight in his own way of life. Like many, he found the religious life the most precious and the most perfect form of human existence. They wasted few regrets over their brothers who turned heretic: they were Judases, and the Whitfords were happy to accept that as a fact. There was a Judas in every congregation.

This is not to say that the bad and the evil justified the existence of the Church, but that the Church did not suffer from the existence of bad individuals, just as Christ had tolerated Judas among his disciples.

What are we to say about the emerging attitudes toward the religious life after considering the evidence presented in the previous chapters? We have examined a mass of evidence, opinions, arguments, and counter-arguments and seen that to everything there was always a second opinion. What does this tell us about sixteenth-century culture? Have we revealed the mentalities of sixteenth-century English people? After tracing the words people wrote in the later Middle Ages and at the beginning of the early modern period, the question arises: Do these texts give us an image of society? I suggest they do. The questions of sin and vice, charity, chastity and sexuality, obedience, poverty, and stability, which we have discussed above, reveal the ways in which early modern people thought.

For the early sixteenth-century English, the world of God was more than important: they hoped to enter the heavenly paradise that awaited them and expected to use their lives to prepare themselves for it. They aimed to pursue a life that was worthy of God, and in one form or another, everyone was supposed to share the yoke of Christ. They knew how they were expected to behave and understood the rules that governed their culture. Religion was an essential part of their mentality with which they could navigate in their world.

In the end, most people had very little to say against the principles the religious life was built upon. As we have seen, the essential structure of monasticism – the three vows (obedience, poverty and chastity) and the question of stability – were not necessarily objectionable ideals to the evangelicals, either. These issues were all parts of their own religious sentiment. What arises from the dialogue is that those who opposed the religious life rejected the notion that the religious life was automatically worthier than life in the world. Yet, why were the religious also sometimes represented as sinful, disobedient, living in luxury, slaves to their flesh, and useless to anyone? This kind of rhetoric speaks of the contest to possess God. It would be foolish to ignore the question of owning God in the early sixteenth-century struggle for power. Seen in this context, the traditional forms of faith and religious in particular were an obvious target of scorn and degradation for their critics. The issues mentioned above were useful weapons in the feud. All parties in the debate agreed that a 'chaste life' was the key to the heart of God, and, similarly, they could renounce flesh or excess wealth, but could argue about the way these ideals should be implemented. Religious debate was a project in which already existing margins and lines were constantly reconstructed and re-enacted.[3]

Because politics created and maintained margins, I have also referred where relevant to what could be called 'politics.' In the sixteenth century, religion and politics could and should not be kept apart. As Michel de Certeau asserts, 'beliefs and practices could confront one another within a *political* space.'[4] Regarding the religious orders, not many evangelicals suggested anything original, and not many even suggested that all religious houses should be suppressed. Religious, however, became pawns in a struggle for (religious) power, and the debate on the religious life thus offers a revealing way into exploring the mentalities of the English.

Where then, one might ask, does the role of anti-clericalism and anti-monasticism (and anti-fraternalism) in the politics of the early sixteenth century lie? Does anti-monasticism have any role in what happened to religious houses? Richard Rex has written that explaining the Reformation is simple 'if anticlericalism, the allegedly long-standing and bitter resentment of the layman against the character and pretensions of the clergy, was really widespread in late medieval England.'[5] Rex is right: it is too simple. I am certain that it is by now obvious that I do not believe in the existence of widespread, systematic sentiments against the clerical order. Here are a few explanations why: monks, nuns, friars, and secular priests were not an alien species; they were all someone's sons, brothers, daughters, aunts, cousins, and uncles. They could even be fathers and mothers. All religious belonged to a family in the world, and they all were part of their communities. Similarly, monasteries recruited locals, and thus their members were known to the surrounding communities. Often it was the families, not religious orders or parishes, that for example educated their sons as priests – it was a popular and respected career, often with good prospects.[6] Why would people systematically despise these members of their families for doing what they wanted them to do in the first place? Regardless of the unavoidable fact that priests occasionally failed, it seems that the English were mainly quite satisfied with their clergy. Whereas there were riots and mobbings of priests in Germany during the most active years of the religious turmoil, there was no such wide-scale violence in England during the reign of Henry VIII; indeed, what risings there were regarding the clergy took place in their defense. The German situation simply cannot explain the English.

Another question related to this issue is the wealth of the clergy and the envy it allegedly incurred. For Dickens, the wealth and powers of the clergy generated anti-clericalism as 'a natural reaction.'[7] There are, however, several points to be considered that speak against this. 'Anti-clericalism' is usually identified as a characteristic of the 'middle classes', whereas the 'lower classes' have not been given a voice. Also, people were used to an extremely hierarchical society, especially so when it came to matters of education and wealth. Could it be a later, a modern, understanding that the poor, the 'deprived,' would experience contempt and envy or even hatred for all those that had more than they did? I do not wish to imply that the English of the sixteenth century were naïve; I rather want to stress that they thought in a different way than we do. They grew up in a society in which there was no discussion of the equality of all people or of democracy. Is it not therefore likely that 'social envy' is a modern sentiment unfamiliar in an age structured in a different manner? Theirs was a society structured by God and the Church, and they lived within a culture of strong hierarchies. I am convinced that people never thought that each and every monk and nun was a bad one; they certainly knew that there were rotten apples in the barrel, but why generalize? If religious were criticized, why would we read that as a complete rejection? Even religious themselves criticized the religious life; they sought to amend things, not to *abolish* the institutions. Monasticism was not an empty ideal; it was for many a living expression of faith.

In sixteenth-century religion, the public expression of faith was essential. It was not enough to believe in private. Vice and sin had to be avoided as strictly as humanly possible, and virtue had to be striven for at every moment. All religion, both lay and religious, aimed at perfecting human life; the goal was the angelic life in heaven. Upon earth, faith was physical: it meant bearing the yoke of Christ in a suitable way. For a layman or a laywoman, the yoke was usually slightly lighter than for a monk or a nun. For a religious, faith was meant to be extremely corporeal: all bodily movement, all everyday activities were strictly controlled. The body was supposed to be punished for its sinfulness. This was done by mortification: by exercises such as fasting, silence, or wearing a hair shirt. A good life in a monastery was still often depicted as the ideal form of human life even though lay virtue was never denigrated as such.

For sixteenth-century English people, obedience was a question of hierarchy and divine will. God was the highest authority and most people were able to accept that it was the king who was his vicar on earth, not the pope. In the religious life disobedience was detested so much that when a monk or a nun entered the religious life he or she had to renounce completely his or her own will. The person who after that acted as the will of the religious was the head of the religious house. Abbots, abbesses and priors were autocrats in their houses, which was what the evangelicals criticized the most when they discussed monastic obedience. Religious houses, in their eyes, were quasi-states within the state, with their own rulers and rules. The religious themselves felt that they obeyed God first and only second his servants on earth. In the end, however, it was the virtue of obedience that opened the monastery gates for the king's commissioners and surrendered the seals to him. The king's claim on them was too formidable to be resisted. Even though the king's claim on religious houses were justified in religious terms and with intentions to reform, dissolutions were politics for money.

Poverty was an ideal which revealed a longing for early Christianity. The evangelicals thought that monastic poverty had become corrupted from its original form. Everyone accepted that poverty was necessary to reach God; the rich had to pay great attention to their salvation. In religious houses, all individuals had to give up their worldly possessions, and if they did not do so, a great hue and cry followed. Everyone seems to have agreed that excess wealth was not good for the Church, but at the same time individuals who donated property to the Church were assured of being rewarded in heaven. People were extremely proud of grand churches, whether they were religious or lay. Decorations had to be lavish in order to please the contemporary eye. The religious could not be exceptions to the prevailing rule; they could not act against the customs of the day. This is what the evangelicals could then easily criticize; it was plain enough that the ideals of the desert fathers were long gone. Criticism against 'corrupt practices' was however much easier than actually doing something concrete about it. Most people did not expect religious to destroy their valuable images and vestments. Moreover, it was ordinary people who had given all the wealth to the religious houses, and continued to donate to them.

The role of the body in religion seems to have been somewhat different for the Catholics and the evangelicals. Whereas the former emphasized religion and the pursuit toward God as a very bodily experience, with concrete bodily feelings, the evangelicals spoke with the same bodily terms but stressed the spiritual aspect of the pursuit. The practical role of the body in religion was not as important for the evangelicals as for those who supported the old faith. For all, however, as said before, Christ's suffering was the most important ideal; bearing the yoke of Christ was the most important duty of Christian life. Chastity was a valid ideal for all people, for religious and their severest critics alike, but the evangelicals did not put so much emphasis on virginity as the religious did; for the evangelicals, marriage was the ideal chaste heaven, for the religious, the religious house. Many still thought it possible to achieve a chaste life in a religious house.

The issues that could be included under the question of stability were met with equal concern by both the Catholics and the evangelicals. Charity, teaching, and preaching were of equal importance for both factions. Only praying actually changed its meaning: the traditional prayers for the soul were in reforming eyes more or less useless. Nonetheless, the evangelicals could not deny the value of the work done in religious houses. After the dissolution many previous critics admitted that religious houses had not been completely evil for the poor. The majority of people thought the religious life justified as religious did carry out visible charitable work.

It should be asked why the people of England so tamely let their monasteries be so severely criticized and then completely dissolved. Several explanations can be found. It seems that defending the Church in itself was not seen as immediately important; averting the danger which the reforming movements represented was regarded as a much more immediate cause. For long the pope's Church was considered so unwavering and omnipotent that it was impossible to envisage its destruction in England. Defending the monasteries in a very active way was not considered necessary because the dissolution, in the end, could not be foreseen either. Furthermore, standing up to a king who actually could legitimize his claim with God's word was extremely dangerous, and ultimately recognized as futile. Those who defended their houses were hanged, drawn and quartered, and their severed heads posted at gates and bridges. The dissolutions were an event that happened; they had to be accepted.

Patricia Cricco proposes that the Reformation was a consequence of the conservative petrification of the Church and its institutions: the Church had not been able to reform itself in time. In Cricco's opinion the English Church failed in devotional terms, and this failure was followed by a popular protest.[8] This claim is rather problematic, as it seems to indicate that monasteries had failed to convince people of their religiosity. Had it been so, there should have been visible more popular religious protest in England at that time; iconoclasm or such other marks of a wide sentiment of dissatisfaction should have been more prominent.

When discussing the reasons for the English Reformation, Ronald Hutton comes to the conclusion that no popular reaction to the actions from above were needed as people were allowed to continue their Catholic traditions: even if this was not possible in church, people performed many of the forbidden rituals themselves, with little interference from the clergy. Thus the transformation from Catholic to Protestant England was smooth.[9] I am ready to support Hutton's view, but wish to stress that his proposition must be widened. It was not only the rituals that people wanted to be left untouched. More important was their thinking, their system of navigation in their culture, and their cultural identity. Cultural change is seldom rapid; events may change at a faster pace than actual ideas. My view here is close to that of Katherine French, who has studied the gendering of parish life in late medieval England. She concludes that it was not the Reformation that brought Christianity, as it were, from rites in a church to individual homes; the process had begun long before the beginning of the sixteenth century.[10] Different ideals and different mentalities are capable of coexistence. If changes in religion were slow and ideas could overlap, it is also essential to keep in mind, as Natalie Davis proposes, that people's attitudes to religion could be flexible. The French artisans she has studied could change their religious position according to political and social pressures.[11] Such a view supports my position: changes were not so drastic that lines could not have been crossed.

In explaining why the reforms were possible, the essence of the matter is that in those issues that were truly important for people, little changed. The ideals of obedience, poverty, chastity, and stability were all integral parts of the late medieval and early sixteenth-century worldview and Christianity. To these ideals the evangelicals did not offer much new nor did they question them. This is why the evangelicals were – at least in the long run, if not during the reign of Henry VIII – successful. The changes Henry made were not solely religious or theological, and because people were not forced to alter their mentalities too suddenly they could – mostly without difficulty – consent to what changes there were. Many people did put up a fight in defense of a particular religious house when the dissolutions were at hand, but there was very little they could do if they wanted to save their heads.

Am I saying that it was the medieval mind that triumphed? Yes. The only reasonable explanation for the peaceful nature of the Reformation is in the *longue durée*. It was not so much the new as the old that presented itself in the Henrician Reformation. It was not so much a period of dramatic change as a period of continuity. Religious were for early sixteenth-century people still representatives of God and a holy form of life. People knew that the religious life was not the only acceptable life in God's eyes, but they understood the special nature of the religious pursuit. People knew religious houses could offer an experience which was exclusive, but for most it was not an evil life.

Notes

Introduction

1 Augustine: *Confessions*, p. 220; Whitford: *Pype*, f. 2, 14v. Jean Leclerq poetically describes the monastery as 'a Jerusalem in anticipation, a place of waiting and of desire, of preparation of that holy city towards which we look with joy.' Leclerq: *Love of Learning*, p. 69. For the monastic moral theology in general, see Vandenbroucke: *Morale monastique*, pp. 13–29.

2 Joyce Appleby, Lynn Hunt and Margaret Jacob attest that historians should aim at more than just preserving, recording, or tracing the past, and suggest that they should 'seek significance, explanation, and meaning'. Appleby, Hunt & Jacob: *Telling the Truth*, p. 257.

3 Virtanen: *Kulttuurihistoria*. See also Chartier: *Cultural History*; Duby: 'Histoire culturelle'.

4 Bynum: 'Wonder', p. 26.

5 Immonen: 'Mennyt nykyisyytenä', p. 29.

6 Here I am especially indebted to Anu Korhonen for the fruitful discussions we have had on the way in which the past can be reconstructed and narrated. See also Jordanova: *Representation*, p. 117. Roger Chartier for example, has spoken of representations as tools for studying the past, but the term becomes futile when we realize that what is left of the past, the sources, were indeed a part of the past and in direct relation to it. It is up to the student to interpret that relation. See Chartier: *On the Edge*, p. 94.

7 Darnton: *Great Cat Massacre*, pp. 4–5, 262; Ginzburg: *Clues*; Ginzburg: 'Making Things Strange'.

8 I have chosen not to focus on symbols because symbolic meanings are already themselves products of interpretation. Symbols can be deciphered from signs after close reading and interpretation. I agree with Chartier, who emphasizes that 'not all signs are symbols'. Chartier: *Cultural History*, pp. 103–4.

9 Obedience was not one of the original vows defined by Gratian, but it was included later. In later monasticism, stability was no longer a vow, but it nevertheless deserves attention because the issues related to it remained important. For more on the vows, see Nilsson: 'Gratian' 1991, p. 149

10 *Summe of the holy scrypture*, f. 38v; Monfasani: 'The Theology of Lorenzo Valla', pp. 10–12; Parish: *The Clerical Marriage*, pp. 141, 151–2, 154–5.

11 In early sixteenth-century religious terms 'sovereign' could also refer to heads other than the monarch.

12 Logan: *Runaway Religious*.

13 There are several excellent recent historiographies of the reformation available, and there is no need to reiterate it here. See for example Bainbridge: *Gilds*; MacCulloch: 'Myth'; Marsh: *Popular Religion*; Walsham: 'Providence'; Willen: 'Revisionism Revised'.

14 Davis: *Society and Culture*, pp. 65–95.

15 Scribner: 'Anticlericalism', p. 147.
16 See, for example, Richmond: *English Gentry*, p. 145.
17 See, for example, Aston: *Faith and Fire*, p. 23; Brady: '"You Hate us Priests"', p. 170; McGrath: *Intellectual Origins*, p. 11.
18 See, for example, Marsh: *Popular Religion*, p. 92.
19 Brigden: 'Youth', pp. 55–6; Gasquet: 'Eve of the Reformation', pp. 114–54; Haigh: *English Reformations*; Collinson: 'England', p. 87; Rex: *Henry VIII*, p. 54.
20 Dickens: *Late Monasticism*, p. 153. See also Jørgensen: *Bauer, Narr und Pfaffe*, p. 66.
21 Szittya: 'Antifraternal Tradition', pp. 4, 192.
22 Braet: 'Cucullus', pp. 161–2.
23 For an analysis of early modern sense of humor, see Korhonen: *Fellows of Infinite Jest*.
24 On the 'birth' of vernacular English devotional literature and its uses in late medieval England, see Hughes: *Pastors and Visionaries*. On early popular print in general, see Halasz: *Marketplace of Print*. The Brigittines were active authors: see, for example, *Here begynneth a deuoute tretyse named the Dyetary of ghostly helthe*. On their literary activities, see Ellis: 'Further Thoughts', pp. 229–43.
25 Diarmaid MacCulloch explains fully why this term is appropriate. MacCulloch: *Thomas Cranmer*, pp. 2–3. See also McClendon: *Quiet Reformation*, p. 69.

Chapter 1

1 Bryson: *From Courtesy to Civility*, p. 64; Rex: 'Crisis', p. 866.
2 This is the way obedience was defined in *Shepherds' Kalendar*, p. 34.
3 Bradshaw: *Holy lyfe*, sig. A2v; see also *Tree & XII. frutes*, sig. 4v.
4 Tunstall: *Sermon*, sig. B6. 'Nabugodonosor the verye chylde of the deuyl'.
5 Tunstall: *Sermon*, sig. B5v.
6 Whitford: *Pype*, sig. 26v. See also Bradshaw: *Holy lyfe*, sig. A2v.
7 Rex: 'Crisis'.
8 He is cited by Bonde: *Pylgrimage*, sig. CC4.
9 *Folowyng of Christ*, f. 17–17v.
10 An account of the abbot's power in the Middle Ages is offered by Wollasch: *Mönchtum des Mittelalters*, pp. 9–52.
11 *Dialoges of creatures moralysed*, sig. C1–2.
12 See, for example, Christine de Pizan: *Medieval Woman's Mirror*, p. 180.
13 *Folowyng of Christ*, f. 17.
14 Bonde: *Pylgrimage*, sig. CC1v. See also *Dyetary of ghostly helthe*.
15 Bonde: *Pylgrimage*, sig. CC4.
16 *Tree & XII. frutes*, f. 4v.
17 Whitford: *Pype*, f. 64, 65, 80–85.
18 Hanawalt: *'Of Good and Ill Repute'*, pp. 70–2. For an interesting discussion on drawing sexual margins, see Puff: 'Localizing Sodomy'.
19 Bonde: *Pylgrimage*, sig. CC2v.
20 Rex: 'Crisis'.
21 This, once again, reaches back to the days of Adam and Eve in Paradise. Sommerville: *Sex and Subjection*, pp. 23–8.

22 Christine de Pizan: *City of Ladies*, p. 119.
23 Davis: *Fiction*, p. 83. See also Davis: *Society and Culture*, pp. 88, 124–51; Orlin: *Private Matters*, p. 104. For a recent account of the ideals of women in early modern England, see Mendelson and Crawford: *Women in Early Modern England*; Pollock: 'Teach Her to Live Under Obedience', pp. 231, 244–9. Another interesting account of female obedience is by David Cressy, who discusses the ceremonial practices and the symbolic nature of wedding flowers in the depiction of female obedience, in Cressy: *Birth, Marriage, and Death*, pp. 363–5.
24 Rex: 'Crisis', p. 866; Sommerville: *Sex and Subjection*, pp. 61, 67–9.
25 On monks as *sponsa Christi*, see below.
26 In the following, my interpretation of the text is deliberately literal. It is not, however, my intention to read the lines as a true account of Lydgate's life but as an account of the notion of a typical conversion.
27 Lydgate: *Testament*, sig. B3v.
28 Ibid., sig. B3v.
29 Ibid., sig. B4.
30 Ibid., sig. B5.
31 For example, Bishop Tunstall stressed the interplay between obedience and pride in his sermon. See Tunstall: *Sermon*, sig. B6. See also More: *Treatise upon the Passion*, p. 9.
32 Vives: *Passions*, pp. 7, 118.
33 More: *Letter to a Monk*, pp. 201, 263, 265; More: *Treatise upon the Passion*, p. 7.
34 More: *Treatise upon the Passion*, p. 7.
35 Whitford: *Pype*, f. 28.
36 *Shepherds' Kalendar*, pp. 34–8.
37 Whitford: *The Pype*, f. 34–5.
38 More: *Treatise upon the Passion*, p. 9.
39 Whitford: *Pype*, f. 59v–60.
40 More: *Dyaloge of Comfort*, p. 80.
41 Davis: *Society and Culture*, p. 79.
42 St German: *Diuision*, f. 3.
43 More: *Dialogue*, p. 423.
44 More: *Dyaloge of Comfort*, p. 129.
45 *Images of a verye Chrysten bysshop*, sig. G6v–G7.
46 Whitford: *Pype*, f. 194v.
47 *Reade me frynde and be not wrothe*, sig. F4v–F5.

Chapter 2

1 Figgis: *Divine Right*, p. 22.
2 Rosenwein: *Negotiating Space*, p. 204.
3 Margaret Beaufort's foundations are famous. Her humanistic ideals were realized in the foundations of new university colleges. Wolsey's dissolutions are famous; in 1518 he had been given the right to inspect religious houses which he was eager to reform, especially the Augustinians and the Benedictines. See Bernard: 'The Piety of Henry VIII', p. 74; Elton: *Reform and Reformation*, p. 94; Gwyn: *King's Cardinal*, pp. 270–1; Hoyle: 'Origins of the Dissolution',

pp. 275–30; McClendon: *Quiet Reformation*, pp. 4–5; Scarisbrick: 'Cardinal Wolsey', p. 45; Scarisbrick: *Reformation*, p. 54; Walker: *John Skelton*, p. 146; Walker: *Plays of Persuasion*, pp. 113–14; Wooding: *Rethinking Catholicism*, pp. 16, 38; Youings: *Dissolution*, p. 27.

4   Baumer: *Early Tudor Theory of Kingship*; Bloch: *Royal Touch*; Burgess: 'Divine Right of Kings'; Febvre: *Problem of Unbelief*; Haas: 'Henry VIII's "Glasse of Truthe"'; Kantorowicz: *King's Two Bodies*; Luscombe: 'State and Nature'; McGrade: 'Somersaulting Sovereignty'; Smith: *Henry VIII*.

5   Sommerville notes that in practice both patriarchal and contractarian theories of power often led to the same conclusion that people could not resist the prince. Patriarchalists, she notes, 'argued that political power was exactly like paternal power: people were born in subjection to their prince' and 'the contractarian theory of government tended to allow the people more power against their prince than did patriarchalism, but there were also conservative contractarians who strictly limited the political community's right to take back its power.' Sommerville: *Sex and Subjection*, pp. 8–9.

6   Augustine: *City of God*, p. 215.

7   Markus: *Saeculum*, p. 95.

8   St Thomas Aquinas: *Summa Theologiae*, p. 416; Copleston: *Aquinas*, pp. 240–1; Oberman; *Dawn of the Reformation*, p. 6.

9   On English kingship, see Smith: *Henry VIII*, pp. 51, 54.

10   Kantorowicz: *King's Two Bodies*, pp. 316, 318, 329.

11   Loach: 'Function of Ceremonial', p. 52.

12   Bloch: *Royal Touch*, pp. 25, 54, 65–6, 93–6; Merriman: *Life and Letters*, Vol. II, p. 53. See also Wooding: *Rethinking Catholicism*, pp. 44, 69.

13   d'Avray: 'Papal Authority', pp. 398–9, 406–7.

14   Oberman: *Dawn of the Reformation*, pp. 12, 32; Goodman: *New Monarchy*, pp. 8–9; Tudor-Craig: 'Henry VIII', p. 183.

15   The mendicant orders took an especially active part in the debate. The Franciscans mainly supported the idea of the pope as the sole authority upon earth; the monarch was dependent on Christ's crown, and on earth that crown was represented only by the pope, Christ's vicar. Dominicans, for their part, were more prepared to rely on secular rulers. They stressed that Christ himself had never possessed secular power and furthermore, he had ordered Christians to give Caesar his due. See Chydenius: *Humanism*, p. 46; Coleman: 'Dominican Political Theory', pp. 196–7, 217–18.

16   Swanson: *Religion and Devotion*, p. 9.

17   See, for example, Dudley: *Tree of Commonwealth*, p. 31, in which Dudley tries to establish a new kind of ideal kingship for Henry VIII.

18   Baumer: *Early Tudor Theory of Kingship*, pp. 31–2.

19   St German: *Answere*, sig. A4.

20   *Secretum secretorum*, pp. 307, 309, 319, 321.

21   *Dialoges of Creatures Moralysed*, sig. Q2.

22   Attreed: 'England's Official Rose', pp. 85–8; Brigden: *London and the Reformation*, pp. 129–33; Elton: *Policy and Police*, pp. 46–7, 59–61; Fox: 'Prophecies and Politics', pp. 78–81, 93; Gunn: 'Accession of Henry VIII', pp. 278–81; Jansen Jaech: '"Prophisies"'; Jansen: *Political Protest*.

23   MacCulloch: *Thomas Cranmer*, p. 227; McClendon: *Quiet Reformation*, p. 82.

24   Wrightson: 'Politics of the Parish'.

25 Haas: 'Henry VIII's "Glasse of Truthe"', p. 352; Haas: 'Martin Luther', pp. 317–18, 325. See also Elton: *Policy and Police*, p. 171; Rex: 'Crisis'.

26 See, for example, *Obedience in Church and State* in which are transcribed three of Stephen Gardiner's treatises, pp. 87–93, 129, 181; Tunstall: *A Sermon*, sig. B5v; Barnes: *Supplycatyon*, p. 4 repeats this message, and Tyndale discussed the matter in his *Obedience of a Christen man*. See also Bradshaw: 'Exile Literature', p. 112. On John Fisher and the question of supremacy, see Bradshaw: 'Bishop John Fisher', p. 15; Chadwick: 'Royal Ecclesiastical Supremacy'; Gogan: 'Fisher's View'; Wooding: *Rethinking Catholicism*, p. 7.

27 Rex: 'Crisis'.

28 Tunstall: *Sermon*, sig. D8v, E1, E2v. The question of obedience to the king remained important even after it was realized that the pope would not attack England. In 1545 Edward Walshe, for example, wrote of people's duty to defend their country in his *Office and duety in fightyng for your countrey*.

29 Early modern people considered words extremely important, of which taking oaths is a good example. For the importance of words, see Flynn: 'Blasphemy,' p. 37.

30 For this discussion, see especially MacCulloch: *Thomas Cranmer*. See also Bernard: 'The Piety of Henry VIII'; Wooding: *Rethinking Catholicism*.

31 MacCulloch: *Thomas Cranmer*, pp. 54–5. The changes can be read for example from the two famous title-pages of the Coverdale Bible of 1535 and the Great Bible of 1539. For an excellent analysis of the contents and symbolism of these two, see King: *Tudor Royal Iconography*, pp. 54–74.

32 See chapter 3 below.

33 Youings: *Dissolution*, p. 32; MacCulloch: *Later Reformation*, p. 4; Stafford: *Queens*, p. 5. For Praemunire, see Houlbrooke: *Church Courts*, p. 8.

34 Heal: 'Henry VIII', p. 278.

35 *Valor Ecclesiasticus* was an economic evaluation of the property of the religious houses.

36 It has been suggested, for example, by G. W. Bernard that Henry's motives for dissolutions could have been lack of sympathy toward religious. This is not very convincing since Henry even refounded houses. See Bernard: 'The Piety of Henry VIII', p. 73; MacCulloch: *Thomas Cranmer*, p. 207. Loach has brought to attention the curious fact that Henry's funeral procession spent a night at Syon. This was, of course, to imitate Edward IV's cortège route to Windsor, but could it be perhaps said that the choice was deliberate – that the location still had some religious meaning? Loach: 'Function of Ceremonial', p. 59.

37 Hoyle: 'Origins of the Dissolutions', p. 297.

Chapter 3

1 *Ye olde god and the newe*, f. 4v.

2 See, for example, Barlow: *Dyaloge*, sig. M3v.

3 Tyndale: *Practyse of Prelates*, sig. D3–4. Typically, nuns are not mentioned here. They did not have enough power – because of their small number and because of their sex was regarded as less important – to be dangerous.

4 *Reade me frynde and be not wrothe*, sig. F1v.

5 Gottfried: *Bury St Edmunds*, p. 244; Greene: *Medieval Monasteries*, p. 174.

6   Dawtry: 'Monasticism in Chesire', especially pp. 70–2.
7   Dobson: *Church and Society*, pp. 58–9. See also Hughes: *Pastors and Visionaries*; Thomson: 'Knightly Piety'. Not enough research has been done on establishing the social backgrounds of the friars in England. This is perhaps due the difficulties in tracing friars in general because of scarce documentation. It must be noted here that entering a mendicant house required much less money than entering a contemplative house. Becoming a friar was a good way to pursue educational interests even for those perhaps not otherwise able to study at university. Regarding nuns, it has been often assumed that they were quite aristocratic, but Barbara Harris has shown that most nuns came from the parish gentry. Harris: 'New Look', pp. 92–4.
8   Rosenwein: *Negotiating Space*, pp. 156, 207.
9   John Nelar vs. the abbot of Tavistock. PRO, STAC 2, Bundle 31, f. 53; the same case can be found in Whytefeld vs the abbot of Tavistock PRO, STAC 2, Bundle 25, f. 60. This case dates, I believe, to the late 1520s or early 1530s.
10   Rosser: 'Sanctuary and Social Negotiation', pp. 59, 76–7, 77–n80.
11   Bradshaw: *Holy lyfe*, sig. E6v.
12   Ibid., sig. H4.
13   Whitford: *Holy instrucyons*, f. 81. 'Haue you euer a regarde vnto your souerene elders, and betters. For vnto them apperteyneth fyrst to speke, & vnto the subiect & inninours to herken to gyue hede, & euer to be obedient vnto theyr byddinges, and to gyue them dye reuerence and honor, eueryche, accordyng vnto hys dignite degre, and vnto theyr merite of sanctite, and holines, & so maye you do vnto god seruice & honour in them, and folowe hys wyll in folowynge theyrs. Make not your selfe mate, & felowe, nor yet verye familier, and homely with your superiours. For after the olde prouerbe. *Nimia familiaritas parit contemptum*. Ouermuche boldnes, maketh a person malepart, and vnreuerent. Shewe a meke, & gentel countenance vnto all persons. And what so euer you do at commaundement: let hyt be done without murmur or grudge, & without styckynge or stoppage, frely wilfully and spedfully. *Hilarem datorem diligit deus*. God loueth a glad gyuer, tha dothe hys duete with glad herte, and cherefull by hauiour. And be you euer loth to take any honors. Preminence, souereynete or superiorite, yf you may auoyde them without the offence of god for doubtles he is in most peril, & ieoperdy, that is in moste hyght rowlme, and dignite.'
14   Whitford: *Pype*, f. 193.
15   Absenteeism means systematic absence from one's house for periods of varying length. The reason for absence could be, for example, guardianship of manors or parishes. This was a common problem, and commonly accepted as difficult to solve.
16   See, for example, Of the vycious life of prelates and other curates, against such. PRO, S.P.6, vol 4, f. 70–2.
17   Skelton: *Complete Poems*, pp. 262–3. See also Sayers: 'Violence', p. 536.
18   <Cromwell to the Abbot of Woburn>, <July, 1533>, Merriman: *Life and Letters of Thomas Cromwell*, Vol I, pp. 362–3.
19   John Abbot of Whitby to Cromwell, LP X, 47; The abbot of Warden to Cromwell, BL, Cotton MS., Cleopatra E IV, f. 163.
20   Giles Huncastle vs. Thomas Hamond, the prior of Combermere. PRO, STAC 2, Bundle 19, f. 158.

21 Dubois: *Aspects de la vie monastique*, p. 206; Hughes: *Pastors and Visionaries*, p. 77.
22 Dickens: *Reformation Studies*, pp. 73, 222.
23 *Starkey's Life and Letters*, p. xx.
24 Information against the Augustinian Friars, London. LP VII, 1670.
25 For Luther's view on monastic obedience, that God's rule should always be put first and man's rule second, see Lohse: *Mönchtum und Reformation*, pp. 219, 257.
26 St German: *Diuision*, f. 41.
27 Ibid.
28 *Reade me frynde and be not wrothe*, sig. F3v–F4.
29 Tyndale: *Obedience*, f. 21.
30 More: *Supplication*, p. 128.
31 *Reade me frynde and be not wrothe*, sig. F3v.
32 Ibid., sig. F3v–F4.
33 The Observants had had a formidable supporter in Lady Margaret Beaufort, Henry's grandmother. Brown: 'Wolsey and Ecclesiastical Order', p. 223; Dowling: 'Humanist Support', p. 52; MacCulloch: *Thomas Cranmer*, p. 144; Schoeck: 'Humanism in England', p. 26; Underwood: 'Politics and Piety', pp. 39–52.
34 Cromwell to Henry VIII. July 23 <1533>. Merriman: *Life and Letters of Thomas Cromwell*, Vol I, pp. 360–2.
35 LP XIII/I, 1043; LP VII, 130. It has been widely accepted that Forest was the Queen's confessor, but Keith Brown and Peter Marshall suggest that there is no direct evidence to prove this link. See Brown: 'Wolsey and Ecclesiastical Order', p. 230 and Marshall: 'Papist as Heretic', p. 354. Whether Forest was confessor or not is not important in our case – it is enough to know that he was on Katherine's side regardless of his role at court.
36 *Three Chapters of Letters*, p. 190.
37 Aston: *Faith and Fire*, pp. 276, 303. This case is also discussed in Duffy: *Stripping of the Altars*, p. 404; MacCulloch: *Thomas Cranmer*, p. 214.
38 Marshall: 'Papist as Heretic', p. 356.
39 John Whalley to Cromwell. BL, Cotton MS., Cleop., E.IV, f. 102.
40 *Life of Syr Thomas More*, pp. 223–4.
41 Bedyl to Cromwell. *Three Chapters of Letters*, pp. 40–1.
42 Knowles: *Religious Orders*, pp. 232–3.
43 Jasper Fyloll to Cromwell. BL, Cotton MS., Cleop. E. IV, f. 35; Injuctions on Carthusians. BL. Cotton MS., Cleop. E. IV, f. 26; Knowles 1961, 235–6. A romantic story was soon attached to these monks. The stepdaughter of Thomas More, Margaret Gigs, bribed her way into the prison and tried to keep the monks alive by feeding them. See Knowles: *Religious Orders*, p. 236; McConica: *English Humanists*, p. 265.
44 Knowles: *Religious Orders*, p. 236.
45 John Husee to Lord Lisle. PRO, SP 1, 92, f. 38.
46 Jasper Fyloll to Cromwell. BL, Cotton MS., Cleop. E. IV., f. 127.
47 There had been exceptions: during the Hundred Years' War the clergy were required to carry defensive armour. See Cullum: 'Clergy', p. 188.
48 Bercé: *Revolt and Revolution*, pp. 167–8; Bush: ' "Up for the Commonweal" ', p. 299; Bush: 'Captain Poverty', pp. 29, 31; Clark: 'Reformation in the Far North', pp. 78–9; Davies: 'Popular Religion', pp. 58–9; Gunn: 'Peers, Commons

and Gentry', 70; Haigh: *Reformation and Resistance*, pp. 118–9, 199; Haigh: *English Reformations*, 147–51; James: *Society*, pp. 204–5; Marsh: *Popular Religion*, p. 19; Moreton: 'Walshingham Conspiracy', pp. 29–30; Scarisbrick: *Reformation*, p. 82.

49   The York Articles. October 1536, in *Humanist Scholarship and Public Order*, pp. 166–7; The Depositions Made by Robert Aske. Answers delivered on April 11, 1537 in *Humanist Scholarship and Public Order*, pp. 219, 220, 221; Harrison: 'Pilgrimage of Grace', pp. 72–3. Davies: 'Popular Religion', pp. 75–6.

50   Elton: *Policy and Police*, pp. 144–8; Moreton: 'Walshingham Conspiracy', pp. 29–32, 41–2.

51   The Abbot of Jervaux and Quondam of Fountains, LP XII/I, 1036; Examination of John Dakyn, LP XII/I, 786. See also Knowles: *Religious Orders*, p. 325.

52   Cromwell's list on the crimes of the monks. BL, Cotton MS., Cleop., E.IV. f. 111b; The Abbot of Sawley LP XII/I, 491. See also Haigh: *Reformation and Resistance*, pp. 129, 132–6; Ridley: *Henry VIII*, p. 289; Cross: 'Monks, Friars', p. 446.

53   At Barlings Abbey LP XI, 805; see also Gunn: 'Peers, Commons and Gentry', pp. 52–6; Knowles: *Religious Orders*, pp. 322–33; Ridley: *Henry VIII*, p. 285; Smith: *Henry VIII*, pp. 83–5; and especially James: *Society*, pp. 204–5.

54   Kelly: 'Thorns on the Tudor Rose', p. 5; Bernard: 'Church of England', p. 184.

55   Sir William Fayrfax to Cromwell, LP XII/I, 192.

56   *Starkey's Life and Letters*, p. liv.

57   Robert Dalivyl's prophecy. BL, Cotton MS., Cleop., E. IV., f. 128b; Against Rebellion, LP XI, 1215; The Monks of Furness, LP XII/I, 841.

58   On the cultural meaning of prophecies, see Thomas: *Religion and the Decline of Magic*, pp. 472–5; Watt: *Secretaries of God*; Jansen: *Dangerous Talk*.

59   The abbot of Croyland to Cromwell. BL. Cotton MS., Cleop. E. IV., f. 57b; John, abbot of Leicester, to Cromwell, LP XIII/I, 989. Giving food as gifts and even as bribes was not unusual in early modern culture. For more on this, see Ben-Amos: 'Gifts and Favors'; Davis: Gift, pp. 57, 143; Pelling: *Common Lot*, p. 41.

60   The commissioners to Cromwell. *Three Chapters of Letters*, pp. 255–6, 257–8.

61   Knowles: *Religious Orders*, pp. 378–9; Smith: *Henry VIII*, pp. 106–7.

62   See the Pamphlet on the execution of the three abbots. PRO, SP 2, Vol. 155, f. 57. See also Martin: 'Leadership', p. 113.

63   E. Hord, Prior of Henton, to his brother Alen Horde, LP XIV/I, 269; See also Elton: *Policy and Police*, pp. 107–8.

64   The abbot of Vale Royal to Cromwell. BL. Harley MS., 604, f. 62.; Katherine Buckley to Cromwell. BL. Cotton Ms. Cleop. E. IV. f. 228. She was not alone; as Yvonne Parrey has written, there were nuns who 'were serious about their vocations and who did not want to abandon them.' See Parrey: ' "Devoted Disciples of Christ" ', p. 242.

65   Cross: 'Community Solidarity', p. 252.

66   Clark: 'Reformation in the Far North'.

67   Harris: 'New Look', p. 111.

68   Clark: 'Reformation and Reaction', p. 296.

69   Knowles: *Religious Orders*, pp. 310–11.

70   Layton to Cromwell. *Three Chapters of Letters*, pp. 47–8.

71   See for example John Wilson on the situation in Mount Grace. PRO, SP 2., 157, f. 90. To my knowledge Elizabeth Barton was the only nun to be

executed, but her earlier death was not in direct connection with religious houses. Her treason, of course, was disobedience as well.

Chapter 4

1 Powell and Cook: *English Historical Facts*, pp. 105; Whiting: *Local Responses*, p. 16.
2 Martin: 'Leadership and Priorities', p. 115.
3 Smith: *Land and Politics*, p. 73. See also Cornwall: *Wealth and Society*, pp. 87–9, 138.
4 Hare: 'Monks as Landlords', p. 94.
5 Keen: *English Society*, p. 262.
6 Keen: *English Society*, pp. 260–1; Swanson: *Church and Society*, pp. 30, 83.
7 Thompson: 'Habendum et Tenendum', p. 201; for the same reason ecclesiastical wealth was separated from lay jurisdiction, see Smith: *Land and Politics*, p. 202.
8 For example, many monasteries were obliged to provide for knights: e.g. Peterborough, which had to supply forty knights to Norwich Castle every year. Only the new orders and houses founded by the king escaped this feudal obligation. See Lawrence: *Medieval Monasticism*, p. 117; MacCulloch: 'Bondmen', pp. 91, 93–4.
9 More on this in Cornwall: *Wealth and Society*, pp. 87–8.
10 LP XIII/II, 88.
11 LP XIII/II, 41.
12 See, for example, Barclay: *Fyfte Eglog*, sig. C3v–C4.
13 Dubois: *Aspects de la vie monastique*, p. 219.
14 Whitford: *Pype*, f. 28v, 172–3.
15 Barclay: *Ecloges*, sig. E4.
16 Swanson: *Religion and Devotion*, p. 105.
17 See, for example, Milis: *Angelic Monks*, p. 79; Hanska: ' "And the Rich Man Also Died" ', pp. 133–7.
18 Bradshaw: *Holy lyfe*, sig. E6–E6v.
19 See for example *Shepherds' Kalendar*, p. 52.
20 Whitford, *Pype*, f. 180v.
21 *Reade me frynde and be not wrothe*, sig. D1.
22 *Summe of the holy scrypture*, f. 48–9.
23 A processe or an Exhortation to tendre the Chargis of the true husbondys. BL., Lansdowne 762, Misc. pieces, f. 5–6.
24 St German: *Answere*, sig. G2–G2v.
25 Anne Hudson's study on his vocabulary has found strong Lollard influences in Fish's text. Hudson: *Premature Reformation*, p. 501. This theme is repeated in St German: *Diuision*, sig. 3.4. St German most probably was far from being a Lollard, so the theme was by no means unique to them. Lollards assimilated into Lutherans quite quickly, and it is very difficult to identify who was Lollard and who a Lutheran in the 1530s.
26 Fish: *Supplicacyon*, pp. 1–2.
27 More: *Supplication*, pp. 130–1, 214–15.
28 Tyndale: *Practyse of Prelates*, sig. D5.

29   See, for example, *Litel treatise ageynste the mutterynge of some papistis in corners*, sig. B6–7.
30   *Reade me frynde and be not wrothe*, sig. E2v, E4.
31   Morison: *Remedy*, pp. 135–6.
32   Morison: *Lamentation*, pp. 92–3.
33   More: *Supplication*, pp. 142–3.
34   For Elizabeth Barton, see Neame: *Holy Maid*; Jansen: *Dangerous Talk*; Watt: *Secretaries of God*.
35   These were usually disputes over land, cattle, or estate in general. The cases studied for the present purposes are from the Star Chamber. See also, for example, Roger Batyn and his wife against the abbot of Tavistock. PRO, STAC 2, vol 3, f. 220–23. Abbot of Chertsey vs. Geoffrey Hovis. PRO, REQ 2, Bundle 1, f. 4. Agnes Swetyng against John Sanford, the Abbot of Coggeshall. PRO, REQ 2, Bundle 2, f. 41. Thomas Dunton vs. Thomas Isaac, the Prior of the Augustinians of Bishop's Lynn. PRO, REQ 2, Bundle 2, f. 112. Whiting: *Local Responses*, pp. 17–19, refers to some cases against religious houses.
36   Anonymous plaintiff vs. Abbot of Bindon, Dave Stephen, William Prouse, George Clyff, John Miller and John Blackmore. PRO, STAC 2, Vol. 1, f. 28–30.
37   Philip Brayne vs. the Abbot of Buckfast. PRO, STAC 2, Vol. 6, f. 281.
38   John Galet vs. the Prior of Leominster. PRO, STAC 2, Bundle 18, f. 53.
39   William Perchey vs. the Prior of Malton, John Jackson, John Gaton, and Thomas Redhead. PRO, STAC 2, Bundle 20, f. 59.
40   William Lokewode and Agnes Lokewode against Robert, Abbot of Sawtry. PRO, STAC 2, Bundle 32, f. 154.
41   Gaunt vs. Abbot of Whitby. PRO, STAC 2, Vol. 16, f. 17–18. On July 13, 1528, pirates from Boulogne robbed Gaunt's ship, which was loaded with household goods. The pirates sold and partly donated that cargo to the defendant at Whitby harbor. Gaunt asked the king to return his cargo, since he was sure that the abbot knew that he was buying from pirates.
42   See Blair: 'Burbeck Marble', pp. 44, 48–54; Cherry: 'Pottery and Tile', pp. 194, 197; Clark: 'Reformation', 77; Dunning: 'Last Days', p. 60; Geddes: 'Iron', pp. 175, 177, 179, 180–1; Holdsworth: *Piper*, p. 26; Lawrence: *Medieval Monasticism*, 111; Marks: 'Window Glass', pp. 277, 282–3; Moore: 'Brick', pp. 211–14, 223; Parsons: 'Stone', pp. 20–1; Ramsay: 'Alabaster', pp. 34–7; Savine: *English Monasteries*, pp. 101, 103, 105, 107, 110–11, 122, 123, 128, 130.
43   About the standards of living in religious houses, see Smith: *Land and Politics*, p. 143; Mullett: *Popular Culture*, p. 50; Dyer: *Standards of Living*, p. 32.
44   They included an understeward of the lands, a surveyor, eight yeomen waiters, and other officers on the following fields: keeper of the garners for corn, carver, usher of the hall, keeper of the swans and a lame and feeble man, as well as a man appointed by the Duke of Suffolk.
45   Dickens: *Late Monasticism*, pp. 71–3.
46   Thomas Warley to Lady Lisle. *Lisle Letters*, p. 277.
47   Becon: 'Jewel of Joy', cited in *Starkey's Life and Letters*, p. lxxvii.
48   Enclosures, evictions, and other oppressions by grantee of estates of dissolved monastery of Syon. *Tudor Economic Documents*, Vol. 1, p. 19.
49   More: *Supplication*, p. 168.
50   Brinklow: *Complaynt of Roderick Mors*, pp. 6, 7, 9–10.
51   *Suplicacion to our moste Soveraigne Lorde*, p. 44.

52 Leyland: *Laboryouse Journey*, sig. A7. Bale wrote an introduction to this work.
53 Hutton: 'English Reformation', p. 93.
54 Ferguson: *Articulate Citizen*, pp. 218–19.

Chapter 5

1 Whitford: *Pype*, f. 191–2.
2 Bainbridge: *Gilds*, p. 68.
3 Whitford: *Pype*, f. 192–192v.
4 St German: *Diuision*, f. 4, 5v–6.
5 Erasmus: *De contemptu mundi*, f. 86v.
6 *Reade me frynde and be not wrothe*, sig. F4v.
7 See, for example, Whiting: *Local Responses*, p. 16.
8 Brightwell [John Frith]: *Pistle*, f. 26–26v; see also f. 27v.
9 *Summe of the holy scrypture*, f. 49.
10 Hawes: *Comforte of louers*, 95; Nevill: *Castell of Pleasure*, f. A5v.
11 BL, Harley MS, 913, f. 3–6, 7–8, 10–12, 57.
12 *Specimens of English Literature*, pp. 2–3. The theme of luxurious glass can also be found in Langland: *Vision*, III, pp. 45–72.
13 Solme: *Lordis flayle*, sig. C6v.
14 *Olde god and the newe*, sig. O6v.
15 *Reade me frynde and be not wrothe*, sig. E6, E8v.
16 For example, the glory of Venice consisted mainly of its beautiful buildings among which were many religious houses. See Guylforde: *Pylgrymage*, f. 5v. Guylforde's secretary probably meant Ragusa (present-day Dubrovnik) when he praised the city's 'glorious houses of Relygyon'; ibid., f. 7.
17 Bradshaw: *Holy lyfe*, sig. Q1v–Q2.
18 Ibid., sig. E2–E2v.
19 Ibid., sig. P6.
20 More: *Apologye*, pp. 31–2.
21 More: *Dialogue*, p. 41.
22 See Bradshaw: *Holy lyfe*, sig. R7v.
23 More: *Debellation*, p. 16.
24 More: *Apologye*, p. 92.
25 *Dialoges of Creatures Moralysed*, sig. T1.
26 LP XIII/II, 903.

Chapter 6

1 Ingram: 'Sexual Manners', p. 93. For many, a human being consisted of the soul and body and also of spirit. Bynum: 'Why All the Fuss', p. 13, clarifies this division: 'even when discussing soul (anima) and body (corpus) as components of person, medieval theologians and philosophers did not discuss anything at all like the Cartesian mind/body problem. Late medieval philosophy used the Aristotelian concept of soul as life principle. Thus both in metaphysics and in embryology there was argument over whether the person had one soul or many. Indeed, dualities or binaries were frequently not at

stake. Many discussions of knowing and seeing used a threefold categorization of body (corpus), spirit (animus or spiritus), and soul (anima) that placed experiencing either sense data or even dreams and visions in corpus or spiritus, not anima.'

2   St Augustin of Abingdon: *Myrrour*, sig. A3v–A4.
3   Basille: *Nosegay*, sig. E2.
4   Whitford: *Pype*, f. 210v.
5   St Augustin of Abingdon: *Myrrour*, sig. A7v.
6   See, for example, Erasmus: *De contemptu mundi*, f. 70v; Whitford: *Pype*, f. 210v; Lydgate: *Stans puer ad mensam*, sig. B1; *Dives and pauper*, f. 183–183v. For a fuller discussion on the fear of bestiality, see Thomas: *Man and the Natural* World, pp. 30–1, 36–40.
7   Baker: 'Feminist Post-Structuralist Engagements', p. 376.
8   Whitford: *Holy instrucyons*, f. 73v, 80v.
9   Whitford: *Pype*, f. 231v.
10   For discussion on the relationship of bestiality and civility in early modern courtesy manuals, see Bryson: *From Courtesy to Civility*, p. 108.
11   Bonde: *Pylgrimage*, sig. CC4v–DD1.
12   Whitford: *Pype*, f. 14v.
13   Bonde: *Pylgrimage*, sig. LL1v–LL2.
14   Whitford: *Pype*, f. 10v.
15   Ibid., f. 42, 206v.
16   Bynum: *Female Body*, pp. 164–5, 170.
17   Whitford: *Pype*, f. 42.
18   Longland: *Sermon 1536*, sig. H2, H2v.
19   It was one of the keywords of early Christianity as well. See Perkins: *Suffering Self*.
20   Whitford: *Pype*, f. 10v.
21   Ibid., f. 19.
22   Jaouën and Semple: 'Body Into Text', p. 1; Ross: '"She Wept and Cried"', p. 51. About the development and history of the ideology of pain, see Perkins: *Suffering Self*, pp. 173–94.
23   Whitford: *Pype*, f. 41v–42.
24   Discussion on the medieval cult of Christ's body has been enthusiastic in recent years. See, for example, Aers and Staley: *Powers of the Holy*; Beckwith: *Christ's Body*; Bynum: *Holy Feast*; Bynum: *Fragmentation*; Bynum: *Jesus as Mother*; Bynum: *Resurrection*.
25   Longland: *Sermon 1536*, sig. F2v.
26   Tunstall: *Sermon 1539*, sig. A7v.
27   On the grotesque body, see Paster: *Body Embarrassed*, pp. 3, 14.
28   More: *Dialogue of Comfort*, p. 67.
29   Ibid., pp. 291–2.
30   Petroff: *Medieval Women's Visionary Literature*, p. 12.
31   Merback: *Thief*, p. 102.
32   De Pizan: *City of Ladies*, pp.125–6.
33   Asceticism was brought to early monasticism by Benedict of Nursia, who was influenced by John Cassian's writings. Jankowski: *Pure Resistance*, pp. 59–60. As important mediators were also martyr saints whose 'shocking tortures may have brought some comfort to men and women experiencing similarly

horrific death beds. Their lurid images adorning churches probably had more to do with fear of bodily mutilation and corruption, which in a society without the sophisticated medical care of the late twentieth century often began long before death. Hermit saints were popular in the inhospitable terrain of the fens, where a number of hermits made their dwellings and may have initiated such devotions.' Bainbridge: *Gilds*, p. 64.

34 Whitford: *Werke for householders*, sig. B1v.
35 Boorde: *Dyetary*, pp. 242–3, 251; Schoenfeldt: *Bodies and Selves*, pp. 1, 25, 32. In this light it is not very surprising that cleanliness was an important issue in etiquette manuals: one had to be clean both inside and outside. Bryson: *From Courtesy to Civility*, p. 101.
36 More: *Apologye*, pp. 115–20; *Life of Syr Thomas More*, pp. 223–4.
37 More: *Letter to Bugenhagen*, p. 93.
38 Ibid., p. 95.
39 Davis: 'Lollardy', p. 42.
40 Whitford: *Holy instrucyons*, f. 68.
41 *Tree & XII. frutes*, f. 14v. For more on this practice, see Klapisch-Zuber: *Women*, pp. 310–29.
42 See, for example, *Here foloweth also a veray necessary Epystle of dyscrecyon; Here foloweth a deuoute treatyse of dyscernynge of spyrytes*.
43 As reported by Barlow: *Dyaloge*, sig. M4.
44 *Olde fayth*, sig. H7. On Luther and the monastic flesh, see Lohse: *Mönchtum und Reformation*, pp. 215–17, 235–8, 288–94.
45 *Reade me frynde and be not wrothe*, sig. D1.
46 See, for example, Duffy: *Stripping of the Altars*, p. 59 and Wogan-Browne: 'Chaste Bodies', p. 30. 'Doing nothing' differed from what we might understand by this. For example, praying or meditation was not idleness. See also Burke: 'Invention of Leisure', for an interesting discussion about the term 'leisure' and issues related to it.
47 Hawes: *Conuersion*, f. 2.
48 Ibid., f. 5v.
49 *Reade me frynde and be not wrothe*, sig. D1.
50 Ibid., sig. D7.
51 *Starkey's Life and Letters*, pp. liiv, lv, lvi.
52 Barlow: *Dyaloge*, sig. M2.
53 Whitford: *Crossrowe*, sig. I5v.
54 George Gyffard to Cromwell, BL, Cotton MS, Cleop. E. IV., f. 213.
55 Whitford: *Holy instrucyons*, f. 69.
56 Bradshaw: *Holy lyfe*, sig. A2.
57 Ibid., sig. A3.
58 Whitford: *Pype*, f. 181.
59 Whitford: *Pype*, f. 208; *Tree & XII frutes*, f. 4v.
60 Whitford: *Pype*, f. 208.
61 O'Day: *Professions*, pp. 54–5.
62 *Summe of the holy scrypture*, f. 38.
63 See for example Hughes: *Pastors and Visionaries*, pp. 253, 258–9.
64 Starkey: *Exhortation*, f. 20.
65 Starkey: *Dialogue*, pp. 42–3.
66 Barlow: *Dyaloge*, sig. O2v–O3, P1–P1v.

Chapter 7

1 Bullough: 'Transvestism'; Crane: 'Clothing'; Cressy: *Travesties*, pp. 92–115.
2 *Olde god & the newe*, sig. L4v.
3 Bradshaw: *Holy lyfe*, sig. E6–E6v.
4 Whitford: *Pype*, f. 210v, 211v, 231v.
5 Ibid., f. 48v.
6 *Summe of the holy scrypture*, f. 47v.
7 Whitford: *Pype*, f. 217–217v.
8 Douglas: *Purity and Danger*.
9 Barlow: *Dyaloge*, sig. N4v.
10 Ibid., sig. N4, O1.
11 Brightwell [John Frith]: *Pistle*, f. 18v.
12 Whitford: *Pype*, f. 47.
13 Ibid., f. 50.
14 Ibid., f. 210v.
15 Barlow: *Dyaloge*, sig. N3v.
16 *Here begynneth the Lyfe of the Blessed Martyr saynte Thomas*, sig. A8.
17 Erasmus: *Pylgremage*, sig. A4; Skelton: *Complete Poems*, pp. 252–3, 256–8, 259, 261–2.
18 *Summe of the holy scrypture*, f. 47v, 48.
19 Whitford: *Pype*, f. 211v.
20 Barlow: *Dyaloge*, sig. N4.
21 He talks of this in *Summe of the holy scrypture*, f. 43, which he had translated.
22 It must be added that shaving the heads of fools, criminals, and prostitutes was a widespread practice. Hanawalt: *'Of Good and Ill Repute'*, pp. 24–7; Korhonen: *Fellows of Infinite Jest*, pp. 154–7.
23 Brightwell [John Frith]: *Pistle*, f. 27, 97v.
24 Solme: *Lordis flayle*, sig. C6v.
25 *Olde god and the newe*, sig. O3–O4.
26 *Reade me frynde and be not wrothe*, sig. E5v–E6, E8v.
27 More: *Supplication*, p. 207.

Chapter 8

1 Bynum: *Holy Feast*, pp. 34–5. On the topic of food in the late medieval world in general, see also Bynum: *Jesus as Mother*; Rubin: *Corpus Christi*; Boureau: 'Sacrality'; Camporesi: *Anatomy*. Research on medieval mysticism has stressed the importance of Christ as food, the role of the host in Christian culture, and the cult of the wounds of Christ. Many have lately discussed the wounds sometimes having sexual overtones in art: Christ's side was sometimes presented as a breast which was suckled. See Bynum: *Holy Feast*, p. 247; Schmitt and Baschet: ' "Sexualité" du Christ', p. 343; Beckwith: *Christ's Body*, p. 58; Petroff: *Body and Soul*, p. 215. See Davis: 'Sacred and the Body Social' for a discussion of the Eucharist in the mid-century Protestant controversy in France. Even if the sexual nature of the cult is considered less important, the cult of Christ's body must be seen as elementary in medieval Christianity, and it still existed in the early sixteenth century. Rubin offers a moderate

view on the cult. See Rubin: *Corpus Christi*, 302–5. See also Longland: *Sermon 1536*, sig. A4–B1.

2 Bynum: *Holy Feast*, p. 201. See also Rubin: *Corpus Christi*, p. 218.

3 Whitford: *Pype*, f. 206v.

4 *Shepherds' Kalendar*, p. 56.

5 St Augustin of Abingdon: *Myrrour*, sig. B2v-B3. Too much drinking could also cause dropsy, fever, aches, gout, abscesses, and toothache. The results of excess is discussed by Boureau: 'Sacrality', pp. 7–8. See also Camporesi: *Anatomy*, p. 78, in which he discusses monks and food.

6 Whitford: *Pype*, f. 207; see also Bonde: *Pylgrimage*, sig. DD1–DD1v. See also Korhonen: *Fellows of Infinite Jest*, pp. 153–4, 247–53.

7 *Robert the Deuyll*, sig. B1. Becoming a fool was not an arbitrary punishment for Robert. The relationship of folly and beastliness was a complex one. Korhonen: *Fellows of Infinite Jest*, p. 253; Pelling: *Common Lot*, p. 47.

8 *Images of a verye Chrysten bysshop*, sig. E4v–E5.

9 Langland: *Vision*, V, p. 175ff. For an analysis of gluttony and its many meanings in culture, see Miller: 'Gluttony'.

10 *Hundred mery talys*, f. xixv–xx.

11 *Reade me frynde and be not wrothe*, sig. B5v, D6v.

12 Ibid., sig. E4–E4v.

13 Ioye: *Subuersion of Moris*, sig. F5–5v.

14 Barclay: *Ship of Fooles*, p. 144.

15 *Reade me frynde and be not wrothe*, sig. E4v.

16 St German: *Diuision*, f. 3.

17 Lindsay: *Popiniay*, sig. E3v.

18 *Reade me frynde and be not wrothe*, sig. E6.

19 St. German: *Diuision*, f. 37v–38.

20 Erasmus: *De contemptu mundi*, f. 22v–23.

21 Barlow: *Dyaloge*, sig. N1v–N2.

22 Ioye: *Subuersion of Moris*, sig. F5v.

23 Starkey: *Dialogue*, pp. 77, 127–8, 131–2, 200–1.

24 LP VII, 1670.

25 Richard Beerley to Cromwell. BL, Cotton MS., Cleop. E. IV., f. 161.

26 The abbot of Warden to Cromwell. BL, Cotton MS., Cleop. E. IV., f. 163.

Chapter 9

1 Whitford, *Pype*, f. 205–205v.

2 Ibid., f. 206–206v.

3 Brown: *Body and Society*, pp. 420–2; Elliott: *Fallen Bodies*, p. 16; Foucault: 'Battle for Chastity', pp. 15–23; Leyser: 'Masculinity in Flux', p. 103.

4 Erasmus: *De contemptu mundi*, f. 21–2.

5 Smith: *Treason*, p. 121; Le Goff: *Medieval Imagination*, p. 98. Crawford posits that masturbation was 'less discussed as a sin in the early modern period than in the eighteenth and nineteenth centuries', Crawford: 'Sexual Knowledge', p. 90, but I propose that when related to religious life, masturbation was quite central when discussing a good life.

6 Cited in Bullough: *Christian Inheritance*, p. 8.

7   Walter: *Spectacle*, sig. C3.
8   Erasmus: *De contemptu mundi*, f. 24.
9   Roper: *Holy Household*, pp. 31–2.
10 More: *Dialogue*, p. 308.
11 Whitford: *Pype*, f. 206v–217v.
12 Eliot: *Bankette*, sig. B8.
13 Whitford: *Pype*, f. 37.
14 Wogan-Browne cites *Ancrene Wisse*: 'In the middle of delights and ease and fleshly comforts, who was ever chaste?', Wogan-Browne: 'Chaste Bodies', p. 29.
15 Bugge: *Virginitas*, pp. 2, 19. The sixteenth century is often seen as a transitional period when the control of sexuality gradually shifted from the Church to lay officials. See Ingram: 'Sexual Manners', p. 88. For an excellent analysis of sexuality, virginity and the body in early Christianity, see Brown: *Body and Society*, especially pp. 147–9, 164, 170.
16 More: *Dialogue*, p. 375.
17 Warner: *Joan of Arc*. Strong-minded women built strong cities, 'of perpetual duration in the world', as Christine de Pizan put it. Christine de Pizan: *City of Ladies*, pp. 10–11.
18 Brown: *Body and Society*, p. 170.
19 Bottomley: *Attitudes to the Body*, p. 145. See also Elliott: *Fallen Bodies*, p. 49.
20 Breitenberg: *Anxious Masculinity*, pp. 97–8; Newman: *From Virile Woman*, p. 28.
21 *XII fruytes of the holy goost*, f. 68, 73.
22 Wogan-Browne: 'Chaste Bodies', p. 26.
23 Hawes: *Pastyme*, p. 46.
24 *XII fruytes of the holy goost*, f. 66v.
25 St Peter of Luxemburg: *Next way to heuen*, sig. A3v.
26 Ruggiero: *Boundaries of Eros*, p. 154.
27 *Deceyte of women*, sig. F3v.
28 Eliot: *Bankette*, sig. A2v.
29 Bradshaw: *Holy lyfe*, sig. C5v.
30 Ibid., sig. C3.
31 Bugge: *Virginitas*; McDannell and Lang: *Heaven*, pp. 103–7; Roberts: 'Stephen Langton', p. 105. Scholars have also argued that there was a third sex, which was between male and female, in a way neutral human being whose religious soul dictated asexuality. Many medievalists have pointed out, however, that women who were virgins and deeply devout actually became men. Brown: *Body and Society*, pp. 78–103, 383; Jankowski: *Pure Resistance*, pp. 61–2. R. N. Swanson talks of 'emasculinity' instead of the third sex. See Swanson: 'Angels Incarnate', p. 161. Barbara Newman sees that the idea that a woman needs to become a man in order to save her soul misogynist. Newman: *From Virile Woman*. Pollock reminds us that in the early modern society, the terms 'male' and 'female' were not necessarily linked to biological sex, since men could display feminine qualities and women masculine traits. Pollock: 'Teach Her', p. 233. This view is in line with what Mary Baker says about genders: 'genders are representations of the cultural mediations people make in their engagements with the world, we live multiple genders. These are evident both in the ways we "know" ourselves in relation to social and cultural categories and in the ways others see us and classify us.' Baker: 'Feminist Post-Structuralist Engagements', p. 375.

32 Bugge: *Virginitas*, p. 33.
33 Johnson: *Equal in Monastic Profession*, p. 235.
34 Crawford: 'Sexual Knowledge', 85.
35 Davis: *Society and Culture*, p. 89.
36 See for example Wogan-Browne: 'Chaste Bodies', p. 26.
37 Mirk: *Festyuall*, f. 25.
38 Whitford, *Pype*, f. 37, sig. 204v. See also Callaghan: 'Castrator's Song', pp. 327–8; Kaartinen: 'Evangelical Eunuchs'; Tougher: 'Images of Effeminate Men'.
39 More: *Dialogue*, p. 312.
40 *Deceyte of women*, sig. B3–B4v.
41 *Tree & XII. frutes*, f. 4.
42 *Articles of oure fayth*, f. 2v.
43 Ibid., f. 4–4v.
44 *Shepherds' Kalendar*, p. 56.
45 Crawford: 'Sexual Knowledge', p. 88.
46 Morison: *Lamentation*, pp. 93–4.
47 I hope to be forgiven for using this anachronistic term here for the lack of a sixteenth-century term for what we would call homosexuality.
48 Bullough: 'Sin against Nature', p. 65; Le Goff: *Medieval Imagination*, pp. 97–8. For a compact outline of the medieval development of the approach to sodomy, see Puff: 'Localizing Sodomy', p. 167.
49 Stafford: 'Queens', p. 11.
50 Sawtry: *Mariage of Preistes*, sig. A6.
51 Puff: 'Localizing Sodomy', pp. 176–7. Interestingly, MacCulloch notes on Thomas Cranmer's difficulties in translating Aristotle's passages on male homosexual incest. Instead of translating, Cranmer both left the notion out and turned the other one into heterosexual activity. MacCulloch: *Thomas Cranmer*, p. 58.
52 Satire on the Church. BL. Stowe, 269, f. 15–16.
53 Bray: 'To Be a Man', p. 158.
54 *The images of a verye Chrysten bysshop*, sig. H1–H1v.
55 Warnicke: 'Sexual Heresy', p. 251. It seems that already before the period considered here the English thought that homosexual practices were a special Italian pastime.
56 Morison: *Lamentation*, pp. 93–4.
57 Ibid., p. 95.
58 *Reade me frynde and be not wrothe*, sig. F6.
59 Ibid., sig. E2.
60 Fish: *Supplicacyon*, p. 6.
61 Ibid., p. 7.
62 More: *Dialogue*, p. 83.
63 Ibid., p. 84.
64 See, for example, Medwall: *Nature*, sig. E4–F1. Fessenden: 'The Convent', p. 452.
65 See, for example, *Hundred mery talys*, in which there is a story of two nuns who came to London for their confession. The young and beautiful nun was given a milder penance by the confessor than the old and ugly one. The comments of the confessor were as follows: 'A fayre yonge lady / with a lusty

gallant / in a pleasaunt herber / in ye mery moneth of May / ye dyd but your kynde. Now byt my trouth god fogyue you & I do'; and for the older nun: 'An olde hore to lye with an olde frere / in the olde cloyster / in the holy tyme of Lent. By cokkys body yf god forgyue the yet wyll I neuer fogyue the.' See *Hundred mery talys*, sig. XVIII. The conclusion is that both nuns were whores and the priest was a whorish hypocrite because he treated these two women unequally. And, of course, there was a friar involved as well.

66 Ioye: *Subuersion of Moris*, sig. G3–G4v.
67 *Images of a verye Chrysten bysshop*, sig. T7v–T8.
68 Ibid., sig. Q6–Q6v, T7v–T8.
69 Johansson: 'Mannen och kvinnan', 43.
70 Wiesner: *Gender, Church, and State*, pp. 45–6. Merry Wiesner discussed later the position of the German women in society after the Reformation, and put emphasis on the deterioration of their situation. See Wiesner: 'Women and the Reformation', pp. 9–15. Diana Willen argues that in England this was not the case, as women would not agree to practice religion confined within the walls of the home. See Willen: 'Women and Religion', pp. 155–6.
71 I believe that the idealized virginity of Queen Elizabeth I can be traced back to the medieval ideal. Virginity still gave a woman special, even magical power. On the idealization of her virginity in literature, see Berry: *Of Chastity and Power*.
72 Davis: *Society and Culture*, esp. pp. 88–93.
73 Fish: *Supplicacyon*, p. 14.
74 *Reade me frynde and be not wrothe*, sig. F6v. Clerical marriage in England has been recently discussed in detail by MacCulloch: *Thomas Cranmer* and Parish: *Clerical Marriage*.
75 Walter: *Spectacle of lovers*, sig. C3.
76 Brightwell [John Frith]: *Pistle*, f. XCVIv.
77 Beguines were not, of course, professed nuns and therefore as laywomen living in suspect communities, were suspect as persons.
78 Bale: *Actes*, f. 18, 75, 77. Alan Stewart considers John Bale rather homophobic. Stewart: *Close Readers*, pp. xv, 41–52. It must also be remembered that Bale himself married when clerical marriage was not legalized in England.
79 Fisher: *Sermon verie fruitfull*, p. 470.
80 Ibid., p. 473.
81 More: *Apologye*, pp. 31, 32, 42, 49, 50, 82; More: *Dialogue*, pp. 165, 346, 348, 360, 366, 375, 378, 426, 434; More: *Letter to Bugenhagen*, pp. 17, 29, 57, 99, 103–5.
82 More: *Apologye*, p. 32; Parish: *Clerical Marriage*, p. 79.
83 More: *Dialogue*, p. 165; Parish: *Clerical Marriage*, pp. 146–7.
84 More: *Dialogue*, p. 346; Parish: *Clerical Marriage*, p. 13.
85 More: *Dialogue*, p. 366, 360.
86 More: *Letter to Bugenhagen*, p. 75.
87 More: *Dialogue*, pp. 308–9; for more on his views of clerical celibacy, see also Parish: ' "Beastly is Their Living" ', pp. 139–44.
88 More: *Dialogue*, p. 370. It is notable that More uses the same imagery as Lutherans. For him, monasteries would become brothels after Lutherans had occupied them. For Lutherans, monasteries were brothels unless reformed.

89   More: *Supplication*, pp. 156–7.
90   *Starkey's Life and Letters*, p. lvii.
91   Merriman: *Life and Letters of Thomas Cromwell*, Vol. II, pp. 25–7.
92   John ap Rice to Cromwell. BL., Cotton MS., Cleop. E. IV. f. 120b.
93   Layton to Cromwell. BL., Cotton MS., Cleop. E. IV. f. 115b.
94   LP VIII, 148.
95   The commissioners to Cromwell. BL., Cotton MS., Cleop. E. IV. f. 114.
96   John Bartelet to Cromwell. BL., Cotton MS., Cleop. E. IV. f. 34b.
97   Layton to Cromwell. BL., Cotton MS., Cleop. E. IV. f. 127.
98   The abbot of Warden to Cromwell. BL., Cotton MS., Cleop. E. IV. f. 163.
99   This kind of political ignoring of nuns was suspected by Thomas More in regard of the question of monastic property which we discussed above.
100 Commissioners to Cromwell. BL., Cotton MS., Cleop. E. IV. f. 209.
101 Commissioners to Cromwell. BL, Cotton MS., Cleop. E. IV. f. 210b.
102 Bedyll to Cromwell, Three Chapters of Letters, pp. 98–100.
103 Crimina comperta. BL., Cotton MS., Cleop. E. IV., f. 147; BL., Lansdowne MS., 988, f. 1. and LP X, 364. The Lichfield list probably dates from the 1530s. One of these lists is used by Harrison. His table shows that incontinence was the crime of 29 monastics in the Lake Counties and sodomy of 24. Sins such as masturbation are included in the numbers. These numbers are not extremely high, and there are some houses in which there are no sexual crimes reported at all. See Haigh: *Reformation and Resistance*, p. 73; Haigh: 'Anticlericalism', p. 392; Harrison: 'Pilgrimage of Grace', p. 18, table 2; Knowles: *Religious Orders*, pp. 41, 296–8; Ridley: *Henry VIII*, p. 255; Smith: *Henry VIII*, pp. 112–3.
104 Johnson: *Equal in Monastic Profession*, pp. 128–9.
105 Marshall: *Catholic Priesthood*, p. 145.
106 Barlow: *Dyaloge*, sig. N1v, N2v.
107 On chaste marriage, see Elliott: *Spiritual Marriage*, p. 4; McNamara: 'Chaste Marriage', pp. 24–33.

Chapter 10

1   Logan: *Runaway Religious*, p. 2.
2   Dubois: *Aspects de la vie monastique*, p. 219. On medieval notion of *vagatio*, see also Renaut: 'Vagabondage et mendicité'.
3   *Shepherds' Kalendar*, p. 54.
4   Cited in Logan: Runaway Religious, p. 6.
5   Foot: 'Early Anglo-Saxon Monastery', p. 50.
6   These nuns, following the Franciscan rule, stand out as a paradox in the mendicant ideal. Whereas the male friars were supposed to work among the people, their sisters were extremely strictly enclosed.
7   Hicks: 'English Minoresses', p. 162.
8   Whitford: *Pype*, f. 38, 211v, 215v.
9   Bradshaw: *Holy lyfe*, sig. Q2–Q2v.
10 Ibid., sig. Q2v.
11 *Folowyng of Christ*, f. 23–23v.
12 *Dialoges of Creatures Moralysed*, sig. E1–2.

13   Ibid.
14   *Exempla* offered insights into what happened to monks who left their houses; they were lured by the Devil and had to live the rest of their miserable lives in sin. *Dialoges of Creatures Moralysed*, sig. E1–F1, S3; Dubois: *Aspects de la vie monastique*, p. 219.
15   Logan: *Runaway Religious*, pp. 147–55.
16   Thomas, Prior of Christchurch, Canterbury to Cromwell, LP XIII/II, 139.
17   Whitford: *Pype*, f. 37.
18   Ibid., f. 38; Kaartinen: 'Evangelical Eunuchs'.
19   More: *Letter to a monk*, pp. 201, 263, 265.
20   *Olde god & the newe*, sig. O3v–O4.
21   St German: *Diuision*, f. 8. John Colet agreed: monks ought to devote themselves to prayer, and fasting, and to chastizing their flesh, and observing rules. See Colet: *Sermon*, p. 301.

Chapter 11

1   Nilsson: 'Gratian', p. 150.
2   Mirk, *Festyuall*, f. 7; *Shepherds' Kalendar*, p. 94; St German: *Power of the Clergye*, sig. A8.
3   Voci: 'Petrarca', pp. 15–16, 55–6, 97–9, 113, 122.
4   Many treatises concentrate on this. See, for example, *Epystle of dyscrecyon*; *Dyscernynge of spyrytes*; Bonde: *Pilgrimage*, sig. FFF5v; Bradshaw: *Holy Life*.
5   Carey: 'Devout Literate Laypeople', pp. 361, 381; Hughes: *Pastors and Visionaries*, pp. 258–9; Swanson: *Religion and Devotion*, p. 106.
6   *Scala perfectionis*, chapters 1.4.–1.8.
7   Secularization here does not mean atheism, but the shift of spiritual matters from the church to the individual.
8   See, for example, Aston: *Faith and Fire*, pp. 19–20; McGrath: *Intellectual Origins*, p. 10; Oberman: *Dawn of the Reformation*, pp. 14–16.
9   Crawford: *Women and Religion*, pp. 22–3. See also Hughes: *Pastors and Visionaries*, p. 348; and for Italian 'open houses' for women, see Gill: 'Open Monasteries'.
10   Duffy: *Stripping of the Altars*, pp. 122, 130.
11   Duffy: *Stripping of the Altars*, p. 142; Johnson: *Equal in Monastic Profession*, p. 259.
12   Oberman: *Dawn of the Reformation*, p. 32.
13   Lehmijoki-Gardner: *Worldly Saints*, pp. 146–56. She regards this process as marginalizing lay women's religious experience; see also pp. 156–62.
14   Tyndale: *Exposition*, f. 40v.
15   St German: *Salem and Bizance*, f. 10v. He adds: 'if preesthode be holle and sounde, the churche flouryssheth: and if it be corrupte, the feyth and vertue of the people fadeth also & vanisheth away'; St German: *Diuision*, f. 6–6v.
16   More: *Debellation*, p. 21.
17   Starkey: *Exhortation*, pp. 74–6; Starkey: *Dialogue*, pp. 149–50.
18   Moone: *Short treatyse of certayne thinges abused In the Popysh Church*, pp. 2–3.
19   Ioye: *Subuersion of Moris*, sig. A8.
20   Brightwell [John Frith]: *Pistle*, f. 9v, 11.

21 Tyndale: *Practyse of Prelates*, sig. B1.

22 Erasmus: *Enchridion*, 23–4.

23 More: *Debellation*, 45–6.

24 Whitford: *Pype*, f. 38v. Boswell: *Kindness of Strangers*, pp. 228, 238–9; Daichman: 'Misconduct', pp. 97–9.

25 Whitford: *Pype*, f. 38v–39.

26 Gwynneth: *Declaracion*, f. 45v.

27 Gwyn: *English Austin Friars*, p. 26; Gardner: 'English Nobility', p. 83.

28 Barron: 'Expansion of Education', p. 237; Gardner: 'English Noblity', p. 80.

29 Grendler: *Schooling*, pp. 7–8, 44, 96–100; Gardner: 'English Nobility', pp. 89, 91; Wunder: *He is the Sun*, p. 44. Claire Cross notes that indeed there is proof that the children of the local gentry were taught at monasteries. Cross: *End of Medieval Monasticism*, p. 15.

30 Gardner: 'English Nobility', pp. 82–7. It has been assumed that especially nunneries were aristocratic institutions, but it seems that Tudor recruitment to them was not aristocratic at all. Cooke: 'English Nuns', p. 290: Harris: 'New Look', p. 93. Curiously enough, in France the aristocracy seems to have held monasticism in high value, and for the high nobility of the sixteenth-century France it seems to have been self-evident that monasteries played an important role as benefices for the daughters in their dynastic strategies. See Baker: 'Female Monasticism'.

31 Power: *Medieval English Nunneries*, pp. 239–41, 245–7.

32 Keen: *English Society*, p. 265. The nuns of Syon put great emphasis on reading and on caring for books. See Hutchison: 'Devotional Reading', esp. pp. 217–19.

33 Johnson: *Equal in Monastic Profession*, p. 147.

34 Barker: 'Caritas Pirckheimer', p. 266; Caritas defended her monastic lifestyle against the attacks of the Lutherans to the very end.

35 *Lisle Letters*. See also Gardner: 'English Nobility', pp. 89–90.

36 *Deuout treatise vpon the Pater noster*, sig. A5v.

37 *Articles of oure fayth*, f. 1.

38 McGrath: *Intellectual Origins*, p. 10.

39 Eisenstein: *Printing Press*, p. 55; Sargent: 'Transmission', pp. 225, 226, 228; Walker: *Plays of Persuasion*, pp. 6–7, 131.

40 See for example John Leland's list of monastic collections in BL, Cotton, Vitellius C. IX. f. 227.

41 Becon: *Early Works*, p. 180.

42 *Epistle that Iohan Sturmius*, sig. C4v.

43 Screech: *Laughter*, pp. 265–70; Wooding: *Rethinking Catholicism*, p. 36.

44 Erasmus: *Enchridion*, p. 30.

45 See, for example, *Hundred mery talys*.

46 *Devout treatise vpon the Pater noster*, sig. A3v–A4.

47 *Summe of the holy scrypture*, sig. 47.

48 St German: *Answere*, sig. G1.

49 Scribner: 'Anticlericalism', p. 152. On the invasion of the new learning, see Cobban: *Medieval English Universities*, p. 244; Dickens: *Reformation and Society*, p. 41; Elton: *Policy and Police*, pp. 25–6; Knowles: *English Religious Orders*, pp. 55–6; Smith: *Henry VIII*, p. 92. No such figures are available concerning the English evangelicals.

50 Fish: *Supplicacyon*, p. 14.

51   *Reade me frynde and be not wrothe*, sig. D8.
52   Barlow: D*yaloge*, sig. M4.
53   *Reade me frynde and be not wrothe*, sig. B5v.
54   More: *Debellation*, 46.
55   For an analysis of these polemics, see Marshall: 'Fear'. See also Hoyle: 'Origin of the Dissolution', pp. 276–7.
56   Susan Brigden also suggests that people really believed that the soul needed prayers to get to heaven. See Brigden: *London and the Reformation*, p. 11.
57   More: *Supplication*, pp. 214–15, 218.
58   *Starkey's Life and Letters*, pp. lv–lvi.
59   Ibid., p. lix.
60   Duggan: 'Fear and Confession', pp. 168–9.
61   Marshall: *Catholic Priesthood*, pp. 14–15, 27, 51.
62   See for example Bilinkoff: 'Confessors', pp. 83–4; Duggan: 'Fear and Confession', pp. 168–9; Reinburg: 'Hearing Lay People's Prayer', pp. 19–39.
63   More: *Dyaloge of Comfort*, pp. 45–6.
64   Whitford: *Pype*, f. 43.
65   Barlow: *Dyaloge*, sig. M4v.
66   Brightwell [John Frith]: *Pistle*, sig. 28. Other points of criticism were masses for the dead, multiplying of holy days and worshipping of relics; sig. 29–29v, 30, 30v.
67   *Reade me frynde and be not wrothe*, sig. F2v. See also *Summe of the holy scrypture*, f. 40.
68   Whitford: *Pype*, f. 44–5.
69   Brightwell [John Frith]: *Pistle*, f. 27.
70   Bradshaw: *Holy lyfe*, sig. N6v.
71   *Summe of the holy scrypture*, f. 38.
72   Ibid., f. 38, 43v.
73   See also Shuger: *Habits of Thought*, p. 121.
74   *Olde god & the newe*, sig. P3v–P4.
75   The York Articles. October 1536, in *Humanist Scholarship and the Public Order*, pp. 166–7. The articles also included demands for the setting aside of the heretic Thomas Cromwell and the heretical bishops.
76   See, for example, Ioye: *Subuersion of Moris*, passim.
77   *Olde fayth*, sig. H6v–H7.
78   Of the vycious life of prelates and other curates, against such. PRO, SP 6., vol. 4, f. 70–2.
79   More: *Letter to a monk*, pp. 279, 281–3. Elsewhere, however, he expressed a feeling that the existence of many orders was not evil; after all there were likewise many different military groups who did not clash with each other. They were all needed. Different orders used their time differently, and their differing lifestyles meant different virtues. More: *Letter to Bugenhagen*, pp. 51–3.
80   *Original & sprynge of all sectes*. One curious text, for example, attacks the Carmelites for changing the original name of their order – they were now friars of St Mary Our Lady. The fact that the Virgin Mary, who never was a nun, could present an example was a cause of great wonder for them; it was said that they should rather be called hell's friars, since they had come so far from the original religion. *Olde god and the newe*, sig. N2v. That different

orders would compete against each other was despised. It was shameful to try religion by habits of different colors and to make clothing more important an issue than honesty or good life. *Epistle that Iohan Sturmius*, sig. C4v–C4.

81 Solme: *Lordis flayle*, sig. D2v.

82 Faking miracles was a popular topic for debate, and the evangelicals wanted to reveal the deception. A famous case was the rood of Boxley Priory, which was burned in February 24, 1538. In the crucifix there was included a mechanism that made the head, lips and eyes move. Finucane: *Miracles and Pilgrims*, p. 156; Aston: *Faith and Fire*, p. 267. The evangelicals took examples like this as serious offences for political use; they were easy to deploy. It must be doubted, however, that there was deception in the minds of all the devout who operated these artefacts – there must have been dozens of such statues in regular use solely in England. At least the Abbot of Hailes was ready to suffer a most shameful death if the powers of the holy blood held in his house would prove to be a deception. It had been rumored that the blood was exchanged for fresh at regular intervals, and that it did not perform true miracles. The abbot denied this and an eighty-year old monk of his, who had taken care of the blood for forty years, swore to this as well. Abbot of Hailes to Cromwell, LP XIII/I, 347. Certainly there were cases in which outright deception for economic gain was involved, but they did not, in the opinion of the devout, blur the fact that there were genuine miracles and that many churches and religious houses held relics that truly had miraculous qualities. For the likes of Thomas More, stories of deception and fraud made little difference. People who produced fake miracles had to be punished, but these individuals did not devalue the real miracles, which he knew to be numerous. More: *Dialogue*, pp. 87–8.

83 More: *Apologye*, p. 44; More: *Debellation*, p. 24.

84 No one has actually been able to classify St German as a Catholic or a Protestant. It seems that he was quite an individual thinker, and a widely read one.

85 More: *Letter to Dorp*, p. 111.

86 More: *Apologye*, pp. 47, 60, 66; More: *Debellation*, pp. 19, 24, 27, 200; More: *Dialogue*, pp. 295–97; More: *Letter to Bugenhagen*, p. 17.

87 More: *Debellation*, p. 21.

88 For the continuing Catholicism, Bossy: *English Catholic Community*; MacCulloch: *The Later Reformation*; Wooding: *Rethinking Catholicism*.

89 See, for example, MacCulloch: *The Later Reformation*; Reinburg: 'Hearing Lay People's Prayer', pp. 33–4.

Chapter 12

1 Lupset: *Treatise of Charitie*, f. 31.

2 Mirk: *Festyuall*, f. 14v.

3 Ibid., f. 10v.

4 Levi: *Inheriting Power*, pp. 36–9; Bennett: 'Conviviality and Charity', pp. 19–24, 26–9. See also Moisà's comment on Bennett: 'Conviviality', pp. 223–34 and Brigden: 'Religion and Social Obligation', pp. 67–112; Heal: *Hospitality*,

pp. 228–41; Dyer: *Standards of Living*, pp. 237–40; Hughes: *Pastors and Visionaries*, pp. 55–7.

5 Haigh: *Reformation and Resistance*, p. 121; Scarisbrick: *Reformation*, pp. 51–2; Dyer: *Standards of Living*, pp. 237–40. See also Swanson: 'Problems of Priesthood', p. 848.

6 *Tree & XII frutes*, f. 7.

7 Pedersen: 'Piety and Charity', p. 40.

8 Bradshaw: *Holy lyfe*, sig. M3.

9 *Regularis Concordia*, p. 64; Johnson: *Prayer*, p. 160; Rosser: *Medieval Westminster*, pp. 298–300, 310. For wills demonstrating this, see the wills of John Long, Stephen Hombull, and Jasper Alleyn in *English Historical Documents*, pp. 1034–1035. Sometimes charity worked the other way round. It must be remembered that there were religious houses which themselves needed alms and help from people, and, for example, some poor young women who wanted to enter a monastery were helped by their townspeople who collecting dowries for these women to help them to enter a religious house. See Bennett: *Conviviality and Charity*, p. 22.

10 Rubin: *Medieval English Medicine*, pp. 172–4, 180; Siraisi: *Medieval and Early Renaissance Medicine*, pp. 25, 38–9.

11 Foot: 'Early Anglo-Saxon Monastery', pp. 50–1; Orme and Webster: *English Hospital*; Rubin: *Medieval English Medicine*, p. 186.

12 The early modern term for this was 'impotent'.

13 *Historical Collections of a Citizen of London*, f. 8–9.

14 Copland: *Hye way to the Spytell hous*, sig. A4v–B1. See also Orme and Webster: *English Hospital*, pp. 151–5.

15 Copland: *Hye way to the Spytell hous*, sig. B2.

16 *Reade me frynde and be not wrothe*, sig. E1–E1v.

17 *Regularis Concordia*, p. 64. Felicity Heal argues that the monasteries usually were hospitable even though they often concentrated on entertaining and feeding their rich visitors. Heal: *Hospitality*, pp. 223, 225, 227, 228–41.

18 Copland: *Hye way to the Spytell hous*, sig. C3.

19 Tyndale: *Mammon*.

20 *Olde god & the newe*, sig. L5.

21 Jasper Fyloll to Cromwell, BL., Cotton MS., Cleop. E. IV., f, 35. Other cases: See Layton and Legh to Cromwell: BL., Cotton MS. Cleop. E. IV., f. 114., For similar accusations in popular literature, see *Reade me frynde and be not wrothe*, sig. E5.

22 *Reade me frynde and be not wrothe*, sig. E7v.

23 Ibid., sig. E7–E7v.

24 Ibid., sig. E2. Anne Hudson has pointed out that the reason why the Lollards clashed with friars was that they felt that the friars harmed the real beggars. Hudson: *Premature Reformation*, p. 351. See also Aston: *Faith and Fire*, pp. 111–12, 116, 125. Wycliffe had blamed them for not helping the poor as they should have done, and even for kidnapping their children. It was claimed that once there had been mercy in the friars' habit but that was over immediately after the times of St Francis. White: *Social Criticism*, pp. 9, 12–13; Lambert: *Medieval Heresy*, p. 338. Possibly the anti-fraternal features in Lollard texts can be explained by their socio-geographical location. The areas concentrated on textile industry were especially keen on Lollardism. These were

the Chilterns, London, parts of north Essex, Bristol, and Coventry. There were also Lollards in the Kent, Gloucestershire, Wiltshire, Berkshire, and Suffolk textile industrial areas. In the Midlands and in the Northern England there were no or far fewer Lollards. See, for example, Haigh: *English Reformations*, p. 53. Because there actually were fewer monasteries in those areas, it can be assumed that those areas were popular preaching areas for the itinerant friars since they preferred the towns to the countryside. The numerous friars could have irritated people with their begging and demands for hospitality.

25 *Reade me frynde and be not wrothe*, sig. F4v.
26 More: *Supplication*, pp. 115–26; More: *Debellation*, p. 53.
27 Barlow: *Dyaloge*, sig. O2v–O3.
28 Ibid., sig. P2, P3–P3v.
29 Charity had to continue, and the lion's share was now with private individuals. Ben-Amos: 'Gifts and Favors', p. 295; Wrightson: 'The Politics of the Parish', pp. 21–2. For an introduction to the organization of poor relief in early modern Europe, see Jütte: *Poverty and Deviance*.

Conclusions

1 This list comes from Lupset: *Treatise of Charitie*, pp. 13–14.
2 More: *Letter to a monk*, p. 291.
3 On margins, see Puff: 'Localizing Sodomy', pp. 171, 179–80.
4 de Certeau: *Mystic Fable*, p. 19.
5 Rex: *Henry VIII*, p. 50.
6 'The Church was an all-embracing institution, and its clergy were everywhere.' There were 50,000 priests in England. Haigh: *English Reformations*, pp. 5–6. See also Marshall: *Catholic Priesthood*, p. 109, and Bainbridge: *Gilds*, pp. 71–5. The monks of Durham, for example, were local in origin; there were few exceptions. See Dobson: *Church and Society*, p. 57. Marsh points out that in Norwich 10 percent of all the sons mentioned in the city's wills were clerics. Marsh: *Popular Religion*, p. 86.
7 Dickens: *Late Monasticism*, p. 173.
8 Cricco: 'Hugh Latimer', p. 22.
9 Hutton: 'English Reformation', p. 115. For a review of explanations presented, see Cameron: *European Reformation*, pp. 293–311.
10 French: 'Maidens' Lights', p. 420.
11 Davis: *Society and Culture*, pp. 3–16.

Bibliography

Manuscript sources

British Library, London (BL)

Cotton Manuscripts, Cleopatra
Cotton Manuscripts, Vitellius
Harley Manuscripts
Lansdowne Manuscripts
Stowe Manuscripts, 141
Stowe Manuscripts, 269

Public Record Office, London (PRO)

Exchequer, Treasury of the Receipt:
Court of Requests, Proceedings 2, Henry VIII (REQ 2)

The Court of Star Chamber:
Star Camber Proceedings 2, Henry VIII (STAC 2)

State Papers 1
State Papers 6 (Theological Tracts)

Primary sources

The A.B.C: set forthe by the Kynges maiestie and his Clergye, and commaunded to be taught through out all his Realme. All other vtterly set a part as the teachers therof tender his graces fauour. (London: Wyllyam Powell, 1545).

Articles of oure fayth. *In this boke is conteyned the Articles of oure fayth. The x. commaundementis. The. vii. works of mercy. The. vii. dedely synnes. The .vii. pryncypall virtues. And the .vii. Sacraments of holy Chirche whiche euery curate is bounde for to declare to his parysshens. .iii. tymes in the yere* (s.l. [1509])

St. Augustin of Abingdon, *The myrrour of the chyrche*. Here foloweth a deuout treatyse conteynynge many goostly medytacyons & instruccions to al maner of people / necessary & confortable to the edyfycacion of the soule & body to the loue & grace of god. (London: Wynkyn de Worde, 1521).

St Augustine, *The Confessions of St. Augustine*. Transl. Pusey, E. B. (New York 1957).

St Augustine, *City of God* (Concerning the City of God against Pagans.) (De Civitate Dei) Transl. by Henry Bettenson (Harmondsworth 1981).

Bale, J. *The Actes of Englysh votaryes*, comprehendynge their vnchast practyses and examples by allages, from the worldes begynnynge to thys present yeare, collected out of their owne legendes and chronycles by Johan Bale (Wesel 1546).

Barclay, A. *Here begynneth the Egloges of Alexander Barclay,* priest, wherof the first thre conteineth the miseries of courters and courtes, of all Princes in generall. The matteir wherof was translated vnto Englysshe by the said Alexander in forme of Dialoges, out of a boke named in latin *Miserie Curialium,* compiled by Eneas Siluius Poete and Oratour; which ofter was Pope of Rome, and named Pius. In the whiche the interloquutors ben Cornix & Coridon (London: Humfrey Powell, 1548).

Barclay, A. *The fyfte Eglog of Alexandre Barclay* of the Cytezen and the Uplondysman (London: Wynkyn de Worde, 1518).

Barclay, A. *The Ship of Fooles,* wherin is shewed the folly of all States, with diuers other workes adioyned vnto the same, very profitable and fruitfull for all men. Translated out of Latin into Englishe by Alexander Barclay Priest (s.l. 1570).

Barlow, W. *A Dyaloge Descrybyng the Orygynall Ground of these Lutheran Saccyons,* and many of theyr abusys / compyled by Syr Wyllyam Barlow chanon. London 1531 (Amsterdam: The English Experience, 1974).

Barnes, R. *A Supplycatyon* unto the most exellent and redoubted prince kinge henrye the eyght (s.l., s.a.)

Basille, T. [Thomas Becon?]: *A pleasaunt newe Nosegaye,* full of many godly and swete floures, lately gathered by Theodore Basille (London: John Gough, 1542).

Becon, T. *The Early Works of Thomas Becon.* Being the Treatises Published by Him in the Reign of King Henry VIII. Ed. Ayre, J. (Cambridge: The Parker Society, 1843).

Bonde, W. Pylgrimage. *Here begynneth a deuout treatyse in Englysshe / called the Pylgrimage of Perfection:* very profitable for all christen people to rede: and in especiall / to all relygious Persons moche necessary (London: Richard Pynson, 1526).

Boorde, A. *Introduction and Dyetary with Barnes in the Defence of the Berde* (London: Early English Text Society Extra Series 10, 1870).

Bradshaw, H. *The holy lyfe.* Here begynneth the holy lyfe and history of saynt Werburge / very frutefull for all christen people to rede (London: Richard Pynson, 1521).

Brightwell, R. [Frith, John] *A pistle to the Christen reader.* The Reuelation of Antchrist. Antithesis / Wherin are compared to geder Christes actes and oure holye father the Popes (Marlborou in Hesse 1529).

Brinklow, H. *Complaynt of Roderyck Mors,* somtyme a gray fryre, unto the parliament houwse of England his natural cuntry: For the redresse of certen wicked lawes, cruel customs, and cruel deeveys, (about A.D. 1542), and The Lamentacyon of a Christen Agaynst the Cytye of London, made by Roderigo Mors (A.D. 1545). Ed. Cowper, J. M. (London: Early English Text Society Extra Series XXII, 1874/1904).

Chaucer, G. *The Canterbury Tales* (Harmondsworth: Penguin, 1951).

Colet, J. *Catechism. A life of John Colet, D. D. with a appendix of some of his English writings.* Ed. Lupton, J. H. (London: George Bell and Sons, 1909).

Colet, J. *The Sermon of Doctor Colet, made to the Conuocation at Paulis.* A life of John Colet, D. D. with a appendix of some of his English writings. Ed. Lupton, J. H. (London: George Bell and Sons, 1909).

Copland, R. *The hye way to the Spytell hous.* Robert Copland (London 1536).

Deceyte of Women. To the instruction and ensample of all men, yonge and Olde, newly corrected (London [1560?]).

A Deuout treatise vpon the Pater noster / made fyrst in latyn by the moost famous doctour mayster Erasmus Roterodamus / and tourned in to englishe by a yong vertuous and well lerned gentylwoman of .xix. yere of age. London: Thomas Berthelet, 1524.

Dialoges of Creatures Moralysed. Applyably and edicatyfly to euery mery and ioucounde mater of late translated out of Latyn into our Englysshe tonge right profitable to the gouernaunce of man (s.l., s.a.).

Dives and pauper. London: Thomas Berthelet, 1536.

Dudley, E. *The Tree of Commonwealth*. Ed. Brodie, D. M. (Cambridge: Cambridge University Press, 1948).

Dyetary of ghostly helthe. Here begynneth a deuoute tretyse named the Dyetary of ghostly helthe (London 1521).

Eliot, T. *The Bankette of Sapience*, compyled by Syr Thomas Eliot knyghte, and newely augmented with dyuerse tytles and sentences (London: Thomas Berthelet, 1539).

English Historical Documents. Vol V. 1485–1558. Ed. Williams, C. H. (London 1971).

The Epistle that Iohan Sturmius, a man of great lerning and iugement, sent to the Cardynalles and prelates, that were chosen and appointed by the Bysshop of Rome, to serche out the abuses of the churche. Translated into englysshe by Rychard Morysine (London: Thomas Berthelet, 1538).

Erasmus of Rotterdam: *Praise of Folly and Letter to Martin Dorp 1515*. Translated by Betty Radice with an introduction and notes by A. H. T. Levi (Harmondsworth: Penguin Books, 1980).

Erasmus of Rotterdam: *De Contemptu Mundi*. Transl. Thomas Paynel (London: Thomas Berthelet, 1533).

Erasmus of Rotterdam: *Enchiridion Militis Christiani*. Ed. O'Donnell, A. M. (Oxford: Early English Text Society Original Series 282, 1981).

Erasmus of Rotterdam: *Ye pylgremage*. A dialoge or communication of two persons, deuysyd and setforthe in the late tonge, by the noble and famose clarke Desiderius Erasmus intituled ye pylgremage of pure deuotyon. Newly translatyd into Englishe (s.l., s.a.).

Fish, S. A Supplicacyon. *Four Supplications*. 1529–1553 A.D. for the Beggers by Simon Fish. A Supplicacion to our moste Soveraigne Lorde Kynge Henry the Eyght (1544 A.D.) A Supplication of the Poore Commons (1546 A.D.) The Decaye of England by the great multitude of shepe (1550–3 A.D.). Ed. Cowper, J. M. (Milllwood, New York: Early English text Society Extra Series 13, 1973).

Fisher, J. Sermon verie fruitfull. *The English Works of John Fisher*. Ed. Mayor, J. E. B. (London: Early English Text Society Extra Series XXVII, 1876/1935).

The folowyng of Christ. Lately translated our of latyn into Englysshe / and newly examyned / corrected / and imprinted (London: Thomas Godfray, 1535).

Gyulforde, R. The Pylgrymage: *This is the begynnynge / and contynuaunce of the Pylgrymage of Sir Richarde Guylforde Knyght* / & controuler vnto our late soueraygne lorde kynge Henry the .vii. And howe he went with seruaunts and company towardes Iherusalem (London: R. Pynson, 1511).

Gwynneth, J. *A Declaracion of State*, Wherin all Heretikes doow leade their liues: And also of their continuall indeuer, and propre fruictes, which beginneth in the .38. chapiter, and so to thende of the woorke (London: Thomas Berthelet, 1536/1554).

Hawes, S. *The Conversyon of Swerers. A Joyfull Medytacyon to all Englonde of Coronacyon of Kynge Henry the Eyght*. [London 1509]. Edinburgh: Abbotsford Club, 1865.

Hawes, S. The Comforte of Louers (1515) *The Minor Poems*. Ed. Gluck, F. W. and A. B. Morgan (London, New York, Toronto: Oxford University Press, 1974).

Hawes, S. *The Pastyme of pleasure* (London 1509).

Here begynneth the Lyfe of the Blessed Martyr saynte Thomas (London: Rycharde Pynson, 1510).

Here foloweth a deuoute treatyse of dyscernynge of spyrytes veray necessary for ghoostly lyuers (London: Henry Pepwell, 1521).

Here foloweth also a veray necessary Epystle of dyscrecyon in styrynges of the soule (London: Henry Pepwell, 1521).

The Historical collections of a Citizen of London in the fifteenth century. Ed. Gairdner, J. (London: Camden Society New Series XVII, 1876).

Hughe, W. *The troubled mans medicine* verye profitabel to be redde of al men wherein they may learne pacyently to suffer all kyndes of aduersitie made & wrytten by Wyllyam Hughe to a frende of his (London: John Herford, 1546).

Humanist Scholarship and Public Order. Two Tracts against the Pilgrimage of Grace by Sir Richard Morison with Historical Annotations and Related Contemporary Documents. Ed. Berkowitz, D. S. (Washington, London and Toronto: Folger Books, 1984).

A Hundred mery talys (London 1526).

Hyckescorner (London 1510).

The Images of a verye Chrysten bysshop / and of a counterfayte bysshop (s.l.: Wyllyam Marshall, [1535]).

Ioye, G. *The Subuersion of Moris false foundacion:* where vpon he sweteth to et faste and shoue vnder his shameles shorit / to underproppe the popis chirche: Made by George Ioye (Emdon: Jacob Aurik, 1534).

Langland, W. *The Vision of Piers Plowman.* A crtial edition of the b-text (London: Everyman, JM Dent, 1995).

Letters and Papers, Foreign and Domestic, of the Reign of Henry VIII. 36 vols. Ed. Brewer, J. S., J. Gardiner, R. H. Brodie (London 1862–1932). (*LP*)

Leyland, J. *The laboryouse Journey & serche of Johan Leylande*, for Englandes Antiquitees, geuen of hym as a new yeares gyfte to kynge henry the VIII. in XXXVII. yeare of his Reygne, with declaracyons enlarged by Johan Bale (London 1549).

Life of Syr Thomas More, Somtymes Lord Chancellour of England by Ro: Ba:. Ed. Waughan, E. and P. E. Hallett (London: Early English Text Society Original Series 222, 1950/1957).

Lindsay, D. *The complaynte and testament of a Popiniay.* which lueth sore wounded and maye not dye. tyll euery man hathe herd what he sayth (London: John Byddell, [1530]/1538).

The Lisle Letters. Ed. Byrne, M. S. (Bungay 1985).

Litel tretise ageynste the mutterynge of some papistis in corners (London 1534).

Longland, J. Sermon 1536. *A sermond spoken before the kynge his maiestie at Grenwiche, vpoon good fryday: the yere of our Lord. M.CCCCC.XXXVI.* by Johan Longlond byshope of Lincolne (London 1536).

Lupset, T. *Treatise of Charitie* (London 1533).

Lydgate, J. Testament. *Here begynneth the testament of Johnn Lydgate monke of Berry:* which he made hymselfe / by his lyfe dayes (London: Richard Pynson, 1515).

Lydgate, J. *Stans puer ad mensam.* Otherwyse called the boke of Norture / newly imprinted and very necessary vnto all youthe (London: John Redman, [1540]).

Medwall, H. Nature. *A goodly interlude of Nature* compyled by mayster Henry Medwall chapleyn to the ryght reuerent father in god Johan Morton sotyme cardynall and archebyshop of Canterbury (s.l., s.a.).

Merriman, R. B.: *Life and Letters of Thomas Cromwell*. 2 Vols. Ed. Merriman, R. B. (London 1902/1968).

Mirk, J. *The festyuall* (London 1528).

Moone, P. *A short treatyse of certayne thinges abused In the Popysh Church, longe used*: But now abolyshed, to our consolation, And Gods word auaunced, the lyght of our saluation (Ippylwyche, s.a.).

More, T.: *The Apologye of Syr Thomas More, Knyght*. Ed. Taft, A. I. (London: The Early English Text Society Original series 180, 1930).

More, T. *A Dialogue Concerning Heresies*. 1531. The Yale Edition of the Complete Works of St. Thomas More. Vol. 6, part 1. Ed. Lawler, T. M. C., G. Marc'Hadour and R. Marius (New Haven and London: Yale University Press, 1981).

More, T. Debellation. *The Debellation of Salem and Bisanze*. The Complete Works of St. Thomas More. Vol. 10. Ed. Guy, J., R. Keen, C. Miller and R. McGygan (New Haven and London: Yale University Press, 1987).

More, T. Dyaloge of Comfort. *A Dialogue of Comfort against Tribulation*. The Yale Edition of the Complete Works of St. Thomas More Vol 12. Ed. Manley, E. and L. Martz (New Haven and London: Yale University Press, 1976).

More, T. Letter to Martin Dorp. *In Defense of Humanism. Letter to Martin Dorp, Letter to the University of Oxford, Letter to Edward Lee, Letter to a Monk with a New Text and Translation of Historia Richardi Tertii*. The Yale Edition of the Complete Works of St. Thomas More. Vol. 15. Ed. Kinney, D. (New Haven and London: Yale University Press, 1986).

More, T. Letter to a Monk. *In Defense of Humanism. Letter to Martin Dorp, Letter to the University of Oxford, Letter to Edward Lee, Letter to a Monk with a New Text and Translation of Historia Richardi Tertii*. The Yale Edition of the Complete Works of St. Thomas More. Vol. 15. Ed. Kinney, D. (New Haven and London: Yale University Press, 1986).

More, T. *Letter to Bugenhagen, Supplication of Souls, Letter Against Frith*. The Yale Edition of the Complete Works of St. Thomas More. Vol. 7. Ed. Manley, F., G. Marc'hadour, R. Marius and C. H. Miller (New Haven and London: Yale University Press, 1990).

More, T. Supplication of Souls. *Letter to Bugenhagen, Supplication of Souls, Letter Against Frith*. The Yale Edition of the Complete Works of St. Thomas More. Vol. 7. Ed. Manley, F., G. Marc'hadour, R. Marius and C. H. Miller (New Haven and London: Yale University Press, 1990).

More, T. *Treatise on the Passion*. The Complete Works of St. Thomas More. Vol. 13. Ed. Haupt, G. E. (New Haven and London: Yale University Press, 1976).

Morison, R.: Lamentation. A Lamentation in Which is Showed what Ruin and Destruction Cometh of Seditious Rebellion. *Humanist Scholarship and Public Order. Two Tracts against the Pilgrimage of Grace by Sir Richard Morison with Historical Annotations and Related Contemporary Documents*. Ed. Berkowitz, D. S. (Washington, London and Toronto: Folger, 1984).

Morison, R. Remedy. A Remedy for Sedition Wherin Are Contained Many Things Concerning the Ture and Loyal Obeisance That Commons Owe unto Their Prince and Sovereign Lord the King. *Humanist Scholarship and Public Order. Two Tracts against the Pilgrimage of Grace by Sir Richard Morison with Historical*

Annotations and Related Contemporary Documents. Ed. Berkowitz, D. S. (Washington, London and Toronto: Folger, 1984).

Nevill, W. *The castell of pleasure. The conueyaunce of a dreme how Desyre went to ye castell of pleasure* / wherin was the fardyn of affeccyon inhabyted by Beaute to whome he amerously expressed his loue / vpon the which supplycacion rose grete stryfe dystputacion / & argument betwene Pyte & Dysdayne (London: Hary Pepwell, 1518).

Obedience in Church & State. Three Political Tracts by Stephen Gardiner. Ed. Janelle, P. (Cambridge 1930).

The Olde fayth. *The olde fayth / and euident probacion our of the holy scripture*, that the christen fayth (which is the right, true, old nad unfoubted fayth) hath endured sens the beginning of the worlde. by Heinrich Bullinger. Transl. by Miles Coverdale (s.l. 1547).

Olde god and the newe. *A worke entytle of of the olde faythe & the newe, of the olde doctryne and ye newe / or orygynall begynnynge of Idolatrye* (London: John Byddell, 1534).

The original & sprynge of all sectes & orders by whome, whan or were they beganne. Translated our of hye Dutch in Englysh [by Miles Coverdale?] James Nicolson for John Gough (London 1537).

St. Peter of Luxemburg, *The boke entytuled the next way to heuen* the whiche in true walkynge or goynge is but thre days Iourney / and to go or walke euery day but thre myles as wytnesseth Moyses who sayth . . . (London: Wynkyn de Worde, 1510).

De Pizan, C. *The Book of the City of Ladies.* Transl. Richards, E. J. (New York: Persea, 1405/1998).

De Pizan, C. *Medieval Woman's Mirror of Honor. The Treasury of the City of Ladies.* Transl. Willard, C. C. Ed. Cosman, M. P. (New York: Bard Hall and Persea, 1405/1989).

Reade me frynde and be not wrothe. For I saye nothynge but the trothe (Wessell: Henry Nycolson, [1526]/1546).

Regularis Concordia. The Monastic Agreement of the Monks and Nuns of the English Nation. Translated from the Latin by Thomas Symons (London: Thomas Nelson and Sons, 1953).

Ricardus, Prior S. Victoris Parisiensis: *Here foloweth a veray deuoute treatyse (named Benyamyn) of the myghtes and vertues of mannes soule* / & of the way to true contemplacyon / compyled by a noble & famous doctoure a man of gtere holynes & deuocyon / named Rycharde of Saynt Vyctor (London 1521).

Robert the Deuyll: *Here begynneth the lyfe of the moast myscheuoust Robert the deuyll whiche was afterwarde called the seruant of god* (London: Wynkyn de Worde [1510]).

St German, C. *A treatyse concerninge the power of the clergye / and the lawes of the realme* (London: Thomas Godfray, 1535).

St German, C. *A treatise concernynge the diuision betwene the spirytualtie and temporaltie* (London: Robert Redman, 1530).

St German, C. *An answere to a letter* (London [1535]).

St German, C. *Salem and Bizance. A dialogue betwixte two englyshe men, wherof one was called Salem, and the other Bizance* (London: Thomas Berthelet, 1533).

Sawtry, I. *The defence of the Mariage of Preistes*: Agenst Steuen Gardiner bisshop of Wynchester / Wylliam Repse bisshop of Norwiche / and agenst all the bisshops and preistes of that false popish secte / with a confutacion of their vnaduysed vowes unaduysedly deffined: wherby they haue so wyckedly separated them whom God cowpled in lawfull mariage (Auryk: Jan Troost, 1541).

Scala perfectionis. *Hereafter foloweth the chapyters of this present volume of Walter Hylton named in latyn (Scala perfectionis) englysshed the ladder of perfeccyon* whiche volume is deuyded in two partyes (s.l., s.a).

Secretum Secretorum. Nine English Versions. Ed. M. A. Manzalaoui. Volume I. Text (Oxford: Early English Text Society 276, 1977).

Shepherds' Kalendar: *Kalendar & Compost of Shepherds.* From the original edition published by Guy Marchant in Paris in the year 1493, and translated into English c. 1518: newly edited for the year 1931. Ed. Heseltine, G. C. (London: Peter Davis, 1930).

Skelton, J. *The Complete poems of John Skelton.* Ed. Henderson, P. (London and Toronto: J. M. Dent and Sons, 1948).

Somer and Wynter. *The debate and stryfe betwene Somer and wynter with the estate present of Man* (London: Lawrence Andrew, 1528).

Solme, T. *Here begynneth a traetys callyde the Lordis flayle* handlyde by the Bushops powre thresshere Thomas Solme (Basyl: Theophyll Emlos, 1540).

Specimens of English Literature. From the 'Ploughman's Crede' to the 'Shepheardes Calender' A.D. 1394–A.D. 1579. With Introduction, notes and glossarial index by W. W. Skeat (Oxford: Oxford University Press, 1930).

Starkey, T. *A Dialogue between Pole and Lupset.* Ed. Mayer, T. F. (London: Camden Fourth Series 37, 1989).

Starkey, T. *An Exhortation to the people, instructing theym to Unitie and obedience* (London s.a.).

Starkey's Life and Letters: England in the Reign of Henry the Eighth, Part I. Starkey's Life and Letters. With an appendix, giving an extract from Sir William Forrest's Pleasaunt Poesye of Princelie Practise 1548. Ed. Herrtage, S. J. (London 1878/ 1927).

A Supplicacion to our moste Soveraigne Lorde... *Four Supplications.* 1529–1553 A.D. for the Beggers by Simon Fish. A Supplicacion to our moste Soveraigne Lorde Kynge Henry the Eyght (1544 A.D.) A Supplication of the Poore Commons (1546 A.D.) The Decaye of England by the great multitude of shepe (1550–3 A.D.). Ed. Cowper, J. M. (Millwood, New York: Early English text Society Extra Series 13, 1973).

The Summe of the holy scrypture, and ordenarye of the chrysten teachynge, the true chrysten fayth by the whiche we are all Iustifyed. And of the vertue of babtym after the teachyng of the Gospell, and of the apostles, with an informacyon how all estates shulde lyue, accordynge to the Gospell. Translated by Simon Fish ([London 1550]).

St Thomas Aquinas: *Summa Theologiae. A concise translation.* Ed. McDermott, T. (London 1989).

Three Chapters of Letters Relating to the Suppression of Monasteries. Ed. Wright, T. (London: Camden Series, 1843–44).

Tracy, Richarde: *The profe and declaration of thys proposition: fayth only iustifieth:* gathered & letsoforthe by Richarde Tracy (s.l. s.a).

The tree & XII. frutes *A deuout treatyse called the tree & XII. frutes of the holy goost* (London: Robert Copland, 1534).

Tudor Economic Documents. Vols I–III. Being Select Documents Illustrating the Economic and Social History of the Tudor England. Ed. Tawney, R. H. and E. Power (London 1924).

Tunstall, C. *A sermon of Cvtbert Byshop of Duresme*, made vpon Palme sondaye laste past, before the maiestie of our souerayne lorde Kynge Henry the .viii. kynge of England & of France, defensor of the fayth, lorde of Irelande, and in erth next vnder Christ supreme heed of the Chruche of England (London: Thomas Berthelet, 1539).

The .xii. fruytes of the holy goost (London: Robert Copland, 1534).

Tyndale, W. Exposition. *An Exposition of St Matthew* (s.l., s.a.).

Tyndale, W. *The obedience of a Christen man and how Christen rulers ought to governe* / where in also (yf thou marke diligently) thou shalt fynde eues to perccave the crafty conveyaunce of all iugglers (Marlborow in the lande of Hesse 1528).

Tyndale, W. *The practyse of Prelates*. Whether the Kinges grace maye be separated from hys quene because she was his brothers wyfe (Marborch 1530).

Tyndale, W. *A treatyse of the iustificacyon by faith only, otherwise called the parable of the wyked Mammon* (London: James Nycolson, 1536).

Vives, J. L. *The Passions of the Soul. The Third Book of* De Anima et Vita. Transl. Noreña, C. G. (Lewiston, Queenstown, Lampeter: The Edwin Mellen Press, 1538/1990).

Walshe, E. *The office and duety in fightyng for our countrey*. Set forth with dyuerse stronge argumentes gathered our of the holy scripture prouynge that the affection to the natiue countrey shulde moche more rule is vs christian then in the Turkes and infidels, who were therin so feruent, as by the hystoriis doth appere (London: Johannes Herford, 1545).

Walter, W. *The spectacle of louers*. Here after foloweth a lytell contrauers dyalogue bytwene loue and councell / with many goodly argumentes of good women and bad / very compendyous to all estates / newly compyled by Wyllyam Walter seruaunt vnto Syr Henry Marnaye Knyght chancellour of the Duchye of Lancastre (London: Wynkyn de Worde, 1530).

Whitford, R. *The crossrowe or A.B.C.* Here done folowe two opuscules or small werkes of saynt Bonauenture / moche necessarie & profytable vnto all christians specyallye vnto religyous persons, put into Englyshe by a brother of Syon Rychard Whytforde (London 1537).

Whitford, R. *A dayly exercyse and experynce of dethe* / gathered and set forth, by a brother of syon Rycharde Whytforde (London 1537).

Whitford, R.: *A dialoge or communicacion bytwene the curate or ghostly father: & tha parochiane or ghostly chyld*. For a due preparacion vnto howselynge. John Waylande (London 1537).

Whitford, R. *Here folowthe dyuers holy instrucyons and teachynges very necessarye for the helth of mannes soule*, newly made and set forth by a late brother of Syon Rychard Whitforde (London: Wyllyam Myddylton, 1541).

Whitford, R. The Pype. *Here begynneth the Pype / or Tonne / of the lyfe of Perfection*. The reason or cause wherof dothe playnely appere in the processe (London: Robert Redman, 1532).

Whitford, R. *A werke for housholders / or for them that haue the guydyng or gouernaunce of any company*. Gadred and set forth by a professed brother of Syon / Richarde Whitforde: and newly corrected and prynted agayne with an addicion of polici for housholding / set forth also by the same brother (London: Robert Redman, [1533]/1537).

Secondary sources

Aers, D. and L. Staley, *The Powers of the Holy. Religion, Politics, and Gender in Late Medieval English Culture* (University Park, PA: The Pennsylvania State University Press, 1996).

Appleby, J., L. Hunt and M. Jacob, *Telling the Truth about History* (New York: W.W. Norton, 1995).

Aston, M. *Faith and Fire. Popular and Unpopular Religion, 1350–1600* (London: Hambledon, 1993).

Attreed, L. 'England's Official Rose: Tudor Concepts of the Middle Ages', *Hermeneutics and Medieval Culture*. Ed. Gallacher, Patrick J. H. Damico (Albany: State University of New York Press, 1989), 85–95.

D'Avray, D. L. 'Papal Authority and Religious Sentiment in the Later Middle Ages', *The Church and Sovereignty c. 590–1918*. Ed. Wood, D. (Oxford: Blackwell, 1991), 393–408.

Bainbridge, V. R. *Gilds in the Medieval Countryside. Social and Religious Change in Cambridgeshire c. 1350–1558* (Woodbridge: Boydell, 1996).

Baker, J. 'Female Monasticism and Family Strategy: The Guises and Saint Pierre de Reims', *The Sixteenth Century Journal*, XXVIII (4, 1997) 1091–1108.

Baker, M. 'Feminist Post-Structuralist Engagements with History', *Rethinking History*, 2 (3, 1998) 371–8.

Barker, P. S. D. 'Caritas Pirckheimer: A female Humanist Confronts the Reformation', *The Sixteenth Century Journal*, XXVI (2, 1995) 159–272.

Barron, C. M. 'The Expansion of Education in Fifteenth-Century London', *The Cloister and the World. Essays in Medieval History in Honour of Barbara Harvey*. Ed. Blair, J. and B. Golding. (Oxford: Clarendon Press, 1996), 219–45.

Baumer, F. L. *The Early Tudor Theory of Kingship* (New York: Russell & Russell, 1966).

Beckwith, S. *Christ's Body. Identity, Culture and Society in Late Medieval Writings* (London: Routledge, 1993).

Ben-Amos, I. K. 'Gifts and Favors: Informal Support in Early Modern England', *The Journal of Modern History*, 72 (2, 2000) 295–338.

Bennett, J. M. 'Conviviality and Charity in Medieval and Early Modern England', *Past & Present*, 134 (1992) 19–41.

Bercé, Y-M. *Revolt and Revolution in Early Modern Europe. An Essay on the History of Political Violence* (Manchester: Manchester University Press, 1987).

Bernard, G. W. 'The Church of England c. 1529–1642', *History*, 75 (244, 1990) 183–206.

Bernard, G. 'The Piety of Henry VIII', *The Education of a Christian Society. Humanism and the Reformation in Britain and the Netherlands*. Ed. Amos, N. S., A. Pettegree and H. van Nierop (Aldershot, Brookfield, Singapore, Sydney: Ashgate, 1999), 62–88.

Berry, P. *Of Chastity and Power. Elizabethan Literature and the Unmarried Queen* (London and New York: Routledge, 1989).

Bilinkoff, J. 'Confessors, Penitents, and the Construction of Identities in Early Modern Avila', *Culture and Identity in Early Modern Europe* (1500–1800). Ed.

Diefendorf, B. B. and C. Hesse (Ann Arbor: The University of Michigan Press, 1993), 83–100.

Blair, J. 'Burbeck Marble', *English Medieval Industries. Craftsmen, Techniques, Products.* Ed. Blair, J. and N. Ramsay (London: Hambledon, 1991), 41–56.

Bloch, M. *The Royal Touch. Sacred Monarchy and Scrofula in England and France* (London: Routledge, 1973).

Bossy, J. *The English Catholic Community* (London: Darton, Longman & Todd, 1975).

Boswell, J. *The Kindness of Strangers. The Abandonment of Children in Western Europe from Late Antiquity to the Renaissance* (Chicago: The University of Chicago Press, 1988).

Bottomley, F. *Attitudes to the Body in Western Christendom* (London: Lepers Books, 1979).

Boureau, A. 'The Sacrality of One's Own Body in the Middle Ages', *Corps Mystique, Corps Sacré. Textual Transfigurations of the Body from the Middle Ages to the Seventeenth Century.* Ed. Jaouën, F. and B. Semple (New Haven: Yale French Studies 86, 1994), 5–17.

Bradshaw, B. 'Bishop John Fisher, 1469–1535: the Man and His Works', *Humanism, Reform and the Reformation. The Career of Bishop John Fisher.* Ed. Bradshaw, B. and E. Duffy (Cambridge: Cambridge University Press, 1989), 1–24.

Bradshaw, C. J. 'The Exile Literature of the Early Reformation: 'Obedience to God and the King'', *The Education of a Christian Society. Humanism and the Reformation in Britain and the Netherlands.* Ed. Amos, N. S., A. Pettegree and H. van Nierop (Aldershot: Ashgate, 1999), 112–30.

Brady, T. A. '"You hate us priests": Anticlericalism, Communalism, and the Control of Women at Strasbourg in the Age of the Reformation', *Anticlericalism in Late Medieval and Early Modern Europe.* Ed. Dykema, P. and H. A. Oberman (Leiden: E. J. Brill, 1993), 167–207.

Braet, Herman. 'Cucullus non facit monachum: of Beasts and Monks in the Old French Renart Romance', *Monks, Nuns, and Friars in Mediaeval Society.* Ed. King, E. B., J. T. Schaefer and W. B. Wadley (Sewanee, Tennessee: The Press of the University of the South, 1989), 161–70.

Bray, A. 'To Be a Man in Early Modern Society. The Curious Case of Michael Wigglesworth', *History Workshop Journal*, 41 (1996) 155–65.

Breitenberg, M. *Anxious Masculinity in Early Modern England* (Cambridge: Cambridge University Press, 1996).

Brigden, S. *London and the Reformation* (Oxford: Clarendon Press, 1989).

Brigden, S. 'Religion and Social Obligation in Early Sixteenth-Century London', *Past & Present*, 103 (1984) 67–112.

Brigden, S. 'Youth and the English Reformation', *Past & Present*, 95 (1982) 37–67.

Brown, K. 'Wolsey and Ecclesiastical Order: The Case of the Franciscan Observants', *Cardinal Wolsey. Church, State and Art.* Ed. Gunn, S. J. and P. G. Lindley (Cambridge: Cambridge University Press, 1991), 219–38.

Brown, P. *The Body and Society. Men, Women, and Sexual Renunciation in Early Christianity* (London: Faber and Faber, 1990).

Bryson, A. *From Courtesy to Civility. Changing Codes of Conduct in Early Modern England* (Oxford: Clarendon Press, 1998).

Bugge, J. *Virginitas: An Essay in the History of a Medieval Ideal* (The Hague: Martinus Nijhoff, 1975).

Bullough, V. L. 'Introduction: The Christian Inheritance', *Sexual Practices & the Medieval Church*. Ed. Bullough, V. L. and J. Brundage (Buffalo: Prometheus, 1982), 1–12.

Bullough, V. L. 'The Sin against Nature and Homosexuality', *Sexual Practices and the Medieval Church*. Ed. Bullough, V. L. and J. Brundage (Buffalo: Prometheus, 1982), 55–71.

Bullough, V. L. 'Transvestism in the Middle Ages', *Sexual Practices & the Medieval Church*. Ed. Bullough, V. L. and J. Brundage (Buffalo: Prometheus, 1982), 43–54.

Burgess, G. 'The Divine Right of Kings Reconsidered', *The English Historical Review*, CVII (425, 1992) 837–61.

Burke, P. 'The Invention of Leisure in Early Modern Europe', *Past & Present* (146, 1995) 136–50.

Bush, M. L. 'Captain Poverty and the Pilgrimage of Grace', *Historical Research*, 65 (156, 1992) 17–36.

Bush, M. L. ' "Up for the Commonweal" The Significance of Tax Grievances in the English Rebellion of 1536', *The English Historical Review*, CVI (419, 1991) 299–318.

Bynum, C. W. 'The Female Body and Religious Practice in the Later Middle Ages', *Fragments for a History of the Human Body*. Part One. Ed. Feher, M., R. Naddaff and N. Tazi (New York: Zone, 1989), 160–219.

Bynum, C. W. *Fragmentation and Redemption. Essays on Gender and the Human Body in Medieval Religion* (New York: Zone Books, 1992).

Bynum, C. W. *Holy Feast and Holy Fast. The Religious Significance of Food to Medieval Women* (Berkeley: University of California Press, 1987).

Bynum, C. W. *Jesus as Mother. Studies in the Spirituality of the High Middle Ages* (Berkeley: University of California Press, 1984).

Bynum, C. W. *The Resurrection of the Body in the Western Christianity, 200–1336* (New York: Columbia University Press, 1995).

Bynum, C. W. 'Why All the Fuss about the Body? A Medievalist's Perspective', *Critical Inquiry*, 22 (1, 1995) 1–33.

Bynum, C. W. 'Wonder', *The American Historical Review*, 102 (1, 1997) 1–26.

Callaghan, D. 'The Castrator's Song: Female Impersonation on the Early Modern Stage', *Journal of Medieval and Early Modern Studies*, 26 (2, 1996) 321–53.

Cameron, E. *The European Reformation* (Oxford: Clarendon Press, 1991).

Camporesi, P. *The Anatomy of the Senses. Natural Symbols in Medieval and Early Modern Italy* (Cambridge: Polity Press, 1994).

Carey, H. M. 'Devout Literate Laypeople and the Pursuit of the Mixed Life in Later Medieval England', *Journal of Religious History*, 14 (4, 1987) 361–81.

De Certeau, M. *The Mystic Fable* (Chicago and London: The University of Chicago Press, 1992).

Chadwick, H. 'Royal Ecclesiastical Supremacy', *Humanism, Reform and the Reformation. The Career of Bishop John Fisher*. Ed. Bradshaw, B. and E. Duffy (Cambridge: Cambridge University Press, 1989), 169–203.

Chartier, R. *Cultural History. Between Practices and Representations* (Cambridge: Polity Press, 1988).

Chartier, R. *On the Edge of the Cliff. History, Language, and Practices* (Baltimore: The Johns Hopkins University Press, 1997).

Cherry, J. 'Pottery and Tile', *English Medieval Industries. Craftsmen, Techniques, Products*. Ed. Blair, J. and N. Ramsay (London: Hambledon, 1991), 189–209.

Chydenius, J. *Humanism in Medieval Concepts of Man and Society* (Helsinki: Societas Scientiarum Fennica, 1985).

Clark, J. G. 'Reformation and Reaction at St Albans Abbey, 1530–58', *The English Historical Review*, CXV (461, 2000) 297–328.

Clark, M. A. 'Reformation in the Far North: Cumbria and the Church, 1500–1571', *Northern History*, XXXII (1996) 75–89.

Cobban, A. B. *The Medieval English Universities: Oxford and Cambridge to c. 1500* (Aldershot: Scolar Press, 1988).

Coleman, J. 'The Dominican Political Theory of John of Paris in its Context', *The Church and Sovereignty c. 590–1918. Essays in Honour of Michael Wilks*. Ed. Wood, D. (Oxford: Blackwell, 1991), 187–223.

Collinson, P. 'England', *The Reformation in National Context*. Ed. Scribner, B., R. Porter and M. Teich (Cambridge: Cambridge University Press, 1994), 80–94.

Cooke, K. 'The English Nuns and the Dissolution', *The Cloister and the World. Essays in Medieval History in Honour of Barbara Harvey*. Ed. Blair, J. and B. Golding (Oxford: Clarendon Press, 1996), 287–301.

Copleston, F. C. *Aquinas* (Harmondsworth: Penguin Books, 1955/1982).

Cornwall, J. C. K. *Wealth and Society in Early Sixteenth Century England* (London: Routledge, 1988).

Crane, S. 'Clothing and Gender Definition: Joan of Arc', *Journal of Medieval and Early Modern Studies*, 26 (2, 1996) 297–320.

Crawford, P. 'Sexual Knowledge in England 1500–1750', *Sexual Knowledge, Sexual Science. The History of Attitudes to Sexuality*. Ed. Porter, R. and M. Teich (Cambridge: Cambridge University Press, 1994), 82–106.

Crawford, P. *Women and Religion in England 1500–1720* (London: Routledge, 1993).

Cressy, D. *Birth, Marriage, and Death. Ritual, Religion, and the Life-Cycle in Tudor and Stuart England* (Oxford: Oxford University Press, 1997).

Cressy, D. *Travesties and Transgressions in Tudor and Stuart England* (Oxford and New York: Oxford University Press, 2000).

Cricco, P. 'Hugh Latimer and Witness', *The Sixteenth Century Journal*, X (1, 1979) 21–34.

Cross, C. 'Community Solidarity Among Yorkshire Religious After the Dissolution', *Monastic Studies. The Continuity of Tradition*. Ed. Loades, J. (Bangor: Headstart History, 1990), 245–54.

Cross, C. *The End of Medieval Monasticism in the East Riding of Yorkshire* (Beverley: East Yorkshire Local History Series. East Yorkshire Local History Society, 1993).

Cross, C. 'Monks, Friars, and the Royal Supremacy in Sixteenth-Century Yorkshire', *The Church and Sovereignty c. 590–1918. Essays in Honour of Michael Wilks*. Ed. Wood, D. (Oxford: Blackwell, 1991), 437–56.

Cullum, P. H. 'Clergy, Masculinity and Transgression in Late Medieval England', *Masculinity in Medieval Europe*. Ed. Hadley, D. M. (London and New York: Longman, 1999), 178–96.

Daichman, G. S. 'Misconduct in the Medieval Nunnery: Fact, Not Fiction', *That Gentle Strength. Historical Perspectives on Women in Christianity*. Ed. Coon, L. L., K. J. Haldane and E. W. Sommer (Charlottesville and London: University Press of Virginia, 1990), 97–117.

Darnton, R. *The Great Cat Massacre and Other Episodes in French Cultural History* (New York: Vintage, 1984).

Davies, C. S. L. 'Popular Religion and the Pilgrimage of Grace', *Order and Disorder in Early Modern England*. Ed. Fletcher, A. and J. Stevenson (Cambridge: Cambridge University Press, 1985), 58–88.

Davis, J. F. 'Lollardy and the Reformation in England', *Archiv für Reformationsgeschichte. Archive for Reformation History*, 73 (1982) 217–37.

Davis, N. Z. *Fiction in the Archives. Pardon Tales and Their Tellers in Sixteenth-Century France* (London: Polity Press, 1987).

Davis, N. Z. *The Gift in Sixteenth-Century France* (Oxford and New York: Oxford University Press, 2000).

Davis, N. Z. 'The Sacred and the Body Social in Sixteenth-Century Lyon', *Past & Present*, 90 (1981) 40–70.

Davis, N. Z. *Society and Culture in Early Modern France* (Stanford: Stanford University Press, 1975).

Dawtry, A. F. 'Monasticism in Chesire 1092–1300', *Monastic Studies. The Continuity of Tradition*. Ed. Loades, J. (Bangor: Headstart History, 1990), 64–74.

Dickens, A. G. *Late Monasticism and the Reformation* (London: Hambledon, 1994).

Dickens, A. G. *Reformation and Society in Sixteenth-Century Europe* (New York: Harcourt Brace Jovanovich, 1975).

Dickens, A. G. *The Reformation Studies* (London: Hambledon, 1982).

Dobson, R. B. *Church and Society in the Medieval North of England* (London: Hambledon Press, 1996).

Douglas, M. *Purity and Danger. An Analysis of the Concepts of Pollution and Taboo* (London: Routledge, 1991).

Dowling, M. 'Humanist Support for Katherine of Aragon', *Historical Research*, 52 (135, 1984) 46–58.

Dubois, J. *Aspects de la vie monastique en France au Moyen Age* (Aldershot: Variorum, 1993).

Duby, G. 'L'histoire culturelle', *Pour une histoire culturelle*. Ed. Rioux, J-P. and J-F. Sirinelli (Paris: Editions du Seuil, 1997), 427–32.

Duffy, E. *The Stripping of the Altars* (New Haven: Yale University Press, 1992).

Duggan, L. G. 'Fear and Confession on the Eve of the Reformation', *Archiv für Reformationsgeschichte. Archive for Reformation History*, 75 (1984) 153–75.

Dunning, R. W. 'The Last Days of Cleeve Abbey', *The Church in Pre-Reformation England. Essays in Honour of F. R. H. du Boulay*. Ed. Barron, C. M. and C. Harper-Bill (Woodbridge: Boydell, 1985), 58–67.

Dyer, C. *Standards of Living in the Later Middle Ages. Social Change in England c. 1200–1520* (Cambridge: Cambridge University Press, 1989).

Eisenstein, E. L. *The Printing Press as an Agent of Change. Communications and Cultural Transformations in Early-Modern Europe* (New York: Cambridge University Press, 1979/1980).

Elliott, D. *Fallen Bodies. Pollution, Sexuality, and Demonology in the Middle Ages* (Philadelphia: University of Pennsylvania Press, 1999).

Elliott, D. *Spiritual Marriage. Sexual Abstinence in Medieval Wedlock* (Princeton, NJ: Princeton University Press, 1993).

Ellis, R. 'Further Thoughts on the Spirituality of Syon Abbey', *Mysticism and Spirituality in Medieval England*. Ed. Pollard, W. F. and R. Boenig (Cambridge: D. S. Brewer, 1997), 219–43.

Elton, G. R. *Policy and Police. The Enforcement of the Reformation in the Age of Thomas Cromwell* (Cambridge: Cambridge University Press, 1972).

Elton, G. R. *Reform and Reformation. England 1509–1558* (Bungay: Edward Arnold, 1989).

Febvre, L. *The Problem of Unbelief in the Sixteenth-Century. The Religion of Rabelais* (Cambridge, Massachusetts: Harvard University Press, 1982).

Ferguson, A. B. *The Articulate Citizen and the English Renaissance* (Durham, N.C.: Duke University Press, 1965).

Fessenden, T. 'The Convent, the Brothel, and the Protestant Woman's Sphere', *Signs*, 25 (2, 2000) 451–78.

Figgis, J. N. *The Divine Right of Kings* (New York: Harper & Row, 1896/1965).

Finucane, R. C. *Miracles and Pilgrims. Popular Beliefs in Medieval England* (Totowa, N.J.: Rowman and Littlefield, 1977).

Flynn, M. 'Blasphemy and the Play of Anger in Sixteenth-Century Spain', *Past & Present*, 149 (1995) 29–56.

Foot, S. 'What was an Early Anglo-Saxon Monastery?', *Monastic Studies. The Continuity of Tradition*. Ed. Loades, J. (Bangor: Headstart History, 1990), 48–57.

Foucault, M. 'The Battle for Chastity', *Western Sexuality. Practice and Precept in Past and Present Times*. Ed. Ariès, P. and A. Béjin (Oxford: Blackwell, 1985), 14–25.

Fox, A. 'Prophecies and Politic in the Reign of Henry VIII', *Reassessing the Henrician Age. Humanism, Politics and Reform 1500–1550*. Ed. Fox, A. and J. Guy (Oxford: Basil Blackwell, 1986), 77–94.

French, K. L., 'Maidens' Lights and Wives' Stores: Women's Parish Guilds in Late Medieval England', *The Sixteenth Century Journal*, XXIX (2, 1998) 399–425.

Gardner, E. J. 'The English Nobility and Monastic Education, c. 1100–1500', *The Cloister and the World. Essays in Medieval History in Honour of Barbara Harvey*. Ed. Blair, J. and B. Golding (Oxford: Clarendon Press, 1996), 80–94.

Gasquet, F. A. *The Eve of the Reformation. Studies in the Religious Life and Thought of the English People in the Period Preceding the Rejection of the Roman Jurisdiction by Henry VIII* (London: John C. Nimmo, 1900).

Geddes, J. 'Iron', *English Medieval Industries. Craftsmen, Techniques, Products*. Ed. Blair, J. and N. Ramsay (London: Hambledon, 1991), 167–88.

Gill, C. 'Open Monasteries for Women in Late Medieval and Early Modern Italy', *The Crannied Wall. Women, Religion, and the Arts in Early Modern Europe*. Ed. Monson, C. A. (Ann Arbor: The University of Michigan Press, 1992), 15–47.

Ginzburg, C. *Clues, Myths, and the Historical Method* (Baltimore: The Johns Hopkins University Press, 1992).

Ginzburg, C. 'Making Things Strange: The Prehistory of a Literary Device', *Representations*, 56 (1996) 8–28.

Gogan, B. 'Fisher's View of the Church', *Humanism, Reform and the Reformation. The Career of Bishop John Fisher*. Ed. Bradshaw, B. and E. Duffy (Cambridge: Cambridge University Press, 1989), 131–54.

Goodman, A. *The New Monarchy. England, 1471–1534* (Oxford: Blackwell, 1988).

Gottfried, R. S. *Bury St Edmunds and the Urban Crisis: 1290–1539* (Princeton: Princeton University Press, 1982).

Greene, J. P. *Medieval Monasteries* (London and New York: Leicester University Press, 1992).

Grendler, P. F. *Schooling in Renaissance Italy. Literacy and Learning, 1300–1600* (Baltimore: The Johns Hopkins University Press, 1989).

Gunn, S. J. 'The Accession of Henry VIII', *Historical Research*, 64 (155, 1991) 278–88.

Gunn, S. J. 'Peers, Commons and Gentry in the Lincolnshire Revolt of 1536', *Past & Present*, 123 (1989) 52–79.

Gwyn, A. *The English Austin Friars in the Time of Wyclif* (London: Oxford University Press, 1940).

Gwyn, P. *The King's Cardinal. The Rise and Fall of Thomas Wolsey* (London: Barrie and Jenkins, 1990).

Haas, S. W. 'Henry VIII's "Glasse of Truthe"', *History*, 64 (1979) 353–62.

Haas, S. W. 'Martin Luther's "Divine Right" Kingship and the Royal Supremacy: Two Tracts from the 1531 Parliament and Convocation of the Clergy', *The Journal of Ecclesiastical History*, 31 (1980) 317–25.

Haigh, C. 'Anticlericalsim and the English Reformation', *History*, 68 (1983) 391–407.

Haigh, C. *English Reformations. Religion, Politics, and Society under the Tudors* (Oxford: Clarendon Press, 1993).

Haigh, C. *Reformation and Resistance in Tudor Lancashire* (Cambridge: Cambridge University Press, 1975).

Halasz, A. *The Marketplace of Print. Pamphlets and the Public Sphere in Early Modern England* (Cambridge: Cambridge University Press, 1997).

Hanawalt, B. A. *'Of Good and Ill Repute'. Gender and Social Control in Medieval England* (New York and Oxford: Oxford University Press, 1998).

Hanska, J. *'And the Rich Man also died; and He was buried in Hell'* – *The Social Ethos in Mendicant Sermons* (Helsinki: The Finnish Historical Society, 1997).

Hare, J. N. 'The Monks as Landlords. The Leasing of the Monastic Demesnes in Southern England', *The Church in Pre-Reformation England. Essays in Honour of F. R. H. du Boulay*. Ed. Barron, C. M., and C. Harper-Bill (Woodbridge: Boydell, 1985), 82–94.

Harris, B. J. 'A New Look at the Reformation: Aristocratic Women and Nunneries, 1450–1540', *Journal of British Studies*, 32 (2, 1993) 89–113.

Harrison, S. M. *The Pilgrimage of Grace in the Lake Counties, 1536–7* (London: Royal Historical Society, 1981).

Heal, F. 'Henry VIII and the Wealth of the English Episcopate', *Archiv für Reformationsgeschichte. Archive for Reformation History*, 66 (1975) 274–99.

Heal, F. *Hospitality in Early Modern England* (Oxford: Clarendon Press, 1990).

Hicks, M. 'The English Minoresses and their Early Benefactors 1281–1367', *Monastic Studies. The Continuity of Tradition*. Ed. Loades, J. (Bangor: Headstart History, 1990), 158–70.

Holdsworth, C. *The Piper and the Tune. Medieval Patrons and Monks* (Reading: University of Reading, 1991).

Houlbrooke, R. *Church Courts and the People During the English Reformation 1520–1570* (Oxford: Oxford University Press, 1979).

Hoyle, R. W. 'The Origins of the Dissolution of the Monasteries', *The Historical Journal*, 38 (2, 1995) 275–305.

Hudson, A. *The Premature Reformation. Wycliffite Texts and Lollard History* (Oxford: Clarendon Press, 1988).

Hughes, J. *Pastors and Visionaries. Religion and Secular Life in Late Medieval Yorkshire* (Woodbridge: Boydell, 1988).

Hutchison, A. M. 'Devotional Reading in the Monastery and in the Late Medieval Household', *De Cella in Seculum. Religious and Secular Life and Devotion in Late Medieval England*. Ed. Sargent, M. G. (Cambridge: D. S. Brewer, 1989), 215–27.

Hutton, R. 'The English Reformation and the Evidence of Folklore', *Past & Present*, 148 (1995) 89–116.

Immonen, K. 'Mennyt nykyisyytenä', *Metodikirja*. Ed. Kaartinen, M. et al. (Turku: Cultural History, University of Turku, 1993), 19–33.

Ingram, M. 'Sexual Manners: The Other Face of Civility in Early Modern England', *Civil Histories. Essays Presented to Sir Keith Thomas*. Ed. Burke, P., B. Harrison and P. Slack (Oxford: Oxford University Press, 2000), 87–109.

James, M. *Society, Politics and Culture. Studies in Early Modern England* (Cambridge: Cambridge University Press, 1986).

Jankowski, T. A. *Pure Resistance. Queer Virginity in Early Modern English Drama* (Philadelphia: University of Pennsylvania Press, 2000).

Jansen, S. L. *Dangerous Talk and Strange Behavior: Women and Popular Resistance to the Reforms of Henry VIII* (Houndmills: Macmillan – now Palgrave Macmillan, 1996).

Jansen, S. L. *Political Protest and Prophecy under Henry VIII* (Woodbridge: Boydell, 1991).

Jansen Jaech, S. L. 'The "Prophisies of Rymour, Beid, and Marlyng": Henry VIII and a Sixteenth-Century Political Prophecy', *The Sixteenth Century Journal*, XVI (3, 1985) 291–300.

Jaouën, F. and B. Semple, 'Editors's Preface: The Body into Text', *Corps Mystique, Corps Sacré. Textual Transfigurations of the Body from the Middle Ages to the Seventeenth Century*. Ed. Jaouën, F. and B. Semple (New Haven: Yale French Studies 86, 1994), 1–4.

Johansson, K. 'Mannen och kvinnan, lusten och äktenskapet. Några tidstypiska tankegångar kring gåtfulla ting', *Jämmerdal och fröjdesal. Kvinnor is stormaktstidens Sverige*. Ed. Österberg, E. (Stockholm: Atlantis, 1997), 27–70.

Johnson, P. D. *Equal in Monastic Profession. Religious Women in Medieval France* (Chicago: The University of Chicago Press, 1991).

Johnson, P. *Prayer, Patronage, and Power. The Abbey of La Trinité, Vendome, 1032–1187* (New York: New York University Press, 1981).

Jordanova, L. 'The Representation of the Family in the Eighteenth Century: A Challenge for Cultural History', *Interpretation and Cultural History*. Ed. Pittock, J. H. and A. Wear (Houndmills and London: Macmillan – now Palgrave Macmillan, 1991), 109–34.

Jørgensen, N. *Bauer, Narr und Pfaffe. Prototypische Figuren und ihre Funktion in der Reformationsliteratur* (Leiden: E. J. Brill, 1988).

Jütte, R. *Poverty and Deviance in Early Modern Europe* (Cambridge: Cambridge University Press, 1994).

Kaartinen, M., 'Evangelical Eunuchs. The Monk's Body in the English Reformation', *Bodies in Evidence. Perspectives on the History of the Body in Early Modern Europe*. Ed. Kaartinen, M. and A. Korhonen (Turku: Cultural History, University of Turku, 1997), 205–39.

Kantorowicz, E. H. *The King's Two Bodies. A Study in Mediaeval Political Theology* (Princeton: Princeton University Press, 1957/1981).

Keen, M. *English Society in the Later Middle Ages* (Harmondsworth: Allen Lane, 1990).

Kelly, T. J. *Thorns on the Tudor Rose. Monks, Rogues, Vagabonds and Sturdy Beggars* (Jackson: University Press of Mississippi, 1977).

King, J. N. *Tudor Royal Iconography. Literature and Art in an Age of Religious Crisis* (Princeton: Princeton University Press, 1989).

Klapisch-Zuber, C. *Women, Family, and Ritual in Renaissance Italy* (Chicago: The University of Chicago Press, 1985).

Knowles, D. *The Religious Orders in England* (Cambridge: Cambridge University Press, 1961).

Korhonen, A. 'Fellows of Infinite Jest. The Fool in Renaissance England' (Turku: Unpublished PhD dissertation, 1999).

Lambert, M. D. *Medieval Heresy. Popular Movements from Bogomil to Hus* (London: Edward Arnold, 1977).

Lawrence, C. H. *Medieval Monasticism. Forms of Religious Life in Western Europe in the Middle Ages* (London: Longman, 1984).

Leclerq, J. *The Love of Learning and the Desire for God* (London: SPCK, 1974).

Le Goff, J. *The Medieval Imagination* (Chicago: The University of Chicago Press, 1988).

Lehmijoki-Gardner, M. *Worldly Saints. Social Interaction of Dominican Penitent Women in Italy, 1200–1500* (Helsinki: The Finnish Historical Society, 1999).

Levi, G. *Inheriting Power. The Story of an Exorcist* (Chicago: The University of Chicago Press, 1988).

Leyser, C. 'Masculinity in Flux: Nocturnal Emission and the Limits of Celibacy in the Early Middle Ages', *Masculinity in Medieval Europe*. Ed. Hadley, D. M. (London and New York: Longman, 1999), 103–20.

Loach, J. 'The Function of Ceremonial in the Reign of Henry VIII', *Past & Present*, 142 (1994) 43–68.

Logan, F. D. *Runaway Religious in Medieval England, c. 1240–1540* (Cambridge: Cambridge University Press, 1996).

Lohse, B. *Mönchtum und Reformation. Luthers Auseinandersetzung mit dem Mönchsideal des Mittelalters* (Göttingen: Vandenhoeck & Ruprecht, 1963).

Luscombe, D. E. 'The State and Nature and the Origin of the State', *The Cambridge History of Later Medieval Philosophy (from the rediscovery of Aristotle to the disintegration of scholasticism 1100–1600)*. Ed. Kretzmann, N., A. Kenny and J. Pinborg (Cambridge: Camridge University Press, 1982), 757–70.

MacCulloch, D. 'Bondmen Under the Tudors', *Law and Government under the Tudors. Essays Presented to Sir Geoffrey Elton on his Retirement*. Ed. Cross, C., D. Loades and J. J. Scarisbrick (Cambridge: Cambridge University Press, 1988), 91–109.

MacCulloch, D. *The Later Reformation in England, 1547–1603* (Houndmills and New York: Palgrave, 1990/2001).

MacCulloch, D. 'The Myth of the English Reformation', *Journal of British Studies*, 30 (1, 1991) 1–19.

MacCulloch, D. *Thomas Cranmer. A Life* (New Haven and London: Yale University Press, 1996).

Marks, R. 'Window Glass', *English Medieval Industries. Craftsmen, Techniques, Products*. Ed. Blair, J. and N. Ramsay (London: Hambledon, 1991), 265–94.

Markus, R. A. *Saeculum: History and Society in the Theology of St Augustine* (Cambridge: Cambridge University Press, 1989).

Marsh, C. *Popular Religion in Sixteenth-Century England. Holding their Peace* (Basingstoke and London: Macmillan – now Palgrave Macmillan, 1998).

Marshall, P. *The Catholic Priesthood and the English Reformation* (Oxford: Clarendon Press, 1994).

Marshall, P. 'Fear, Purgatory and Polemic in Reformation England', *Fear in Early Modern Society*. Ed. Naphy, W. G. and P. Roberts (Manchester and New York: Manchester University Press, 1997), 150–66.

Marshall, P. 'Papist as Heretic: the Burning of John Forest, 1538', *The Historical Journal*, 41 (2, 1998) 351–74.

Martin, J. 'Leadership and Priorities in Reading during the Reformation', *The Reformation in English Towns 1500–1640*. Ed. Collinson, P. and J. Craig (Basingstoke and London: Macmillan – now Palgrave Macmillan, 1998), 113–29.

McClendon, M. C. *The Quiet Reformation. Magistrates and the Emergence of Protestantism in Tudor England* (Stanford: Stanford University Press, 1999).

McConica, J. K. *English Humanists and Reformation Politics under Henry VIII and Edward VI* (London: Oxford University Press, 1965).

McDannell, C. and B. Lang. *Heaven. A History* (New Haven: Yale University Press, 1988/1990).

McGrade, A. S. 'Somersaulting Sovereignty: A Note on Reciprocal Lordship and Servitude in Wyclif', *The Church and Sovereignty c. 590–1918. Essays in Honour of Michael Wilks*. Ed. Wood, D. (Oxford: Blackwell, 1991), 261–8.

McGrath, A. *The Intellectual Origins of the European Reformation* (Oxford: Blackwell, 1987).

McNamara, J. A. 'Chaste Marriage and Clerical Celibacy', *Sexual Practices & the Medieval Church*. Ed. Bullough, V. L. and J. Brundage (Buffalo: Prometheus, 1982), 22–33.

Mendelson, S. and C. Crawford, *Women in Early Modern England 1550–1720* (Oxford: Clarendon Press, 1998).

Merback, M. B. *The Thief, the Cross and the Wheel. Pain and Spectacle of Punishment in Medieval and Renaissance Europe* (London: Reaktion Books, 1999).

Merriman, R. B. *Life and Letters of Thomas Cromwell. Vols I–II* (London: Oxford University Press, 1968).

Milis, L. J. R. *Angelic Monks and Earthly Men. Monasticism and its Meaning to Medieval Society* (Woodbrigde: Boydell, 1992).

Miller, W. I. 'Gluttony', *Representations*, 60 (1997) 92–122.

Moisà, M. 'Conviviality and Charity in Medieval and Early Modern England', *Past & Present*, 154 (1997) 223–34.

Monfasani, J. 'The Theology of Lorenzo Valla', *Humanism and Early Modern Philosophy*. Ed. Kraye, J. and M. W. F. Stone (London and New York: Routledge, 2000), 1–23.

Moore, N. J. 'Brick', *English Medieval Industries. Craftsmen, Techniques, Products*. Ed. Blair, J. and N. Ramsay (London: Hambledon, 1991), 211–36.

Moreton, C. E. 'The Walshingham Conspiracy of 1537', *Historical Research*, LXIII (150, 1990) 29–43.

Mullett, M. *Popular Culture and Popular Protest in Late Medieval and Early Modern Europe* (London: Croom Helm, 1987).

Neame, A. *The Holy Maid of Kent. The Life of Elizabeth Barton, 1506–1534* (London: Hodder and Stoughton, 1971).

Newman, B. *From Virile Woman to WomanChrist. Studies in Medieval Religion and Literature* (Philadelphia: University of Pennsylvania Press, 1995).

Nilsson, B. 'Gratian: On Entry into the Monastery', *In Quest of the Kingdom. Ten Papers on Medieval Monastic Spirituality.* Ed. Härdelin, A. (Uppsala: Bibliotheca Theologiae Practicae, 1991), 135–55.

Oberman, H. A. *The Dawn of the Reformation. Essays in Late Medieval and Early Reformation Thought* (Edinburgh: T & T Clark, 1986).

O'Day, R. *The Professions in Early Modern England, 1450–1800: Servants of the Commonweal* (Harlow, London, New York: Longman, 2000).

Orlin, L. C. *Private Matters and Public Culture in Post-Reformation England* (Ithaca and London: Cornell University Press, 1994).

Orme, N. and M. Webster. *The English Hospital 1070–1570* (New Haven: Yale University Press, 1995).

Parish, H. ' "Beastly is their living and their doctrine" Celibacy and Theological Corruption in English Reformation Polemic', *Protestant History and Identity in Sixteenth-Century Europe. Vol. 1: The Medieval Inheritance.* Ed. Gordon, B. (Aldershot: Scolar Press, 1996), 138–52.

Parish, H. L. *Clerical Marriage and the English Reformation. Precedent Policy and Practice* (Aldershot, Burlington, Singapore, Sydney: Ashgate, 2000).

Parrey, Y. ' "Devoted disciples of Christ": Early Sixteenth-Century Religious Life in the Nunnery at Amesbury', *Historical Research*, LXVII (164, 1994) 240–8.

Parsons, D. 'Stone', *English Medieval Industries. Craftsmen, Techniques, Products.* Ed. Blair, J. and N. Ramsay (London: Hambledon, 1991), 1–27.

Paster, G. K. *The Body Embarrassed. Drama and the Disciplines of Shame in Early Modern England* (Ithaca: Cornell University Press, 1993).

Pedersen, S. 'Piety and Charity in the Painted Glass of Late Medieval York', *Northern History*, XXXVI (1, 2000) 33–42.

Pelling, M. *The Common Lot. Sickness, Medical Occupations and the Urban Poor in Early Modern England* (London and New York: Longman, 1998).

Perkins, J. *The Suffering Self. Pain and Narrative Representation in the Early Christian Era* (London: Routledge, 1995).

Petroff, E. A. *Medieval Women's Visionary Literature* (New York and Oxford: Oxford University Press, 1986).

Pollock, L. ' "Teach Her to Live Under Obedience": the Making of Women in the Upper Ranks of Early Modern England', *Continuity and Change*, 4 (2, 1989) 231–58.

Powell, K. and C. Cook. *English Historical Facts 1485–1603* (London: Macmillan – now Palgrave Macmillan, 1977).

Power, E. *Medieval English Nunneries, c. 1275–1535* (Cambridge: Cambridge University Press, 1922).

Puff, H. 'Localizing Sodomy: The "Priest and Sodomite" in Pre-Reformation Germany and Switzerland', *Journal of the History of Sexuality*, 8 (2, 1997) 165–95.

Ramsay, N. 'Alabaster', *English Medieval Industries. Craftsmen, Techniques, Products.* Ed. Blair, J. and N. Ramsay (London: Hambledon, 1991), 29–40.

Reinburg, V. 'Hearing Lay People's Prayer', *Culture and Identity in Early Modern Europe (1500–1800).* Ed. Diefendorf, B. B. and C. Hesse (Ann Arbor: The University of Michigan Press, 1993), 19–39.

Renaut, M-H. Vagabondage et mendicité. Délits périmés, réalité quotidienne', *Revue Historique*, CCXCVIII (2, 1998) 287–322.

Rex, R. 'The Crisis of Obedience: God's Word and Henry's Reformation', *The Historical Journal*, 39 (4, 1996) 863–94.

Rex, R. *Henry VIII and the English Reformation* (Houndmills: Macmillan – now Palgrave Macmillan, 1993).

Richmond, C. 'The English Gentry and Religion c. 1500', *Religious Belief and Ecclesiastical Careers in Late Medieval England*. Proceedings of the Conference Held at Strawberry Hill, Easter 1989. Ed. Harper-Bill, C. (Woodbridge: Boydell, 1991), 121–50.

Ridley, J. *Henry VIII* (London: Constable, 1984).

Roberts, P. B. 'Stephen Langton's "Sermo de Virginibus"', *Women of the Medieval World. Essays in Honor of John H. Mundy*. Ed. Kirshner, J. and S. F. Wemple (Oxford: Blackwell, 1985), 103–18.

Roper, L. *The Holy Household. Women and Morals in Reformation Augsburg* (Oxford: Clarendon Press, 1989).

Rosenwein, B. H. *Negotiating Space. Power, Restraint, and Privileges of Immunity in Early Medieval Europe* (Manchester: Manchester University Press, 1999).

Ross, E. ' "She Wept and Cried Right Loud for Sorrow and for Pain" – Suffering, the Spiritual Journey, and Women's Experience in Late Medieval Mysticism', *Maps of Flesh and Light. The Religious Experience of Medieval Women Mystics*. Ed. Wiethaus, U. (Syracuse: Syracuse University Press, 1993), 45–59.

Rosser, G. *Medieval Westminster 1200–1540* (Oxford: Clarendon Press, 1989).

Rosser, G. 'Sanctuary and Social Negotiation in Medieval England', *The Cloister and the World. Essays in Medieval History in Honour of Barbara Harvey*. Ed. Blair, J. and B. Golding (Oxford: Clarendon Press, 1996), 57–79.

Rubin, M. *Corpus Christi. The Eucharist in Late Medieval Culture* (Cambridge: Cambridge University Press, 1991).

Rubin, S. *Medieval English Medicine* (London & Vancouver: David & Charles Newton Abbot, 1974).

Ruggiero, G. *The Boundaries of Eros. Sex Crime and Sexuality in Renaissance Venice* (New York: Oxford University Press, 1985/1989).

Sargent, M. G. 'The Transmission by the English Carthusians of Some Late Medieval Spiritual Writings', *The Journal of Ecclesiastical History*, 27 (1976) 225–40.

Savine, A. *English Monasteries on the Eve of the Dissolution* (Oxford: Clarendon, 1909).

Sayers, J. 'Violence in the Medieval Cloister', *The Journal of Ecclesiastical History*, 41 (1990) 533–42.

Scarisbrick, J. J. 'Cardinal Wolsey and the Common Weal', *Wealth and Power in Tudor England. Essays presented to S. T. Bindoff*. Ed. Ives, E. W., R. J. Knecht and J. J. Scarisbrick (London: The Athlone Press, 1978), 45–67.

Scarisbrick, J. J. *The Reformation and the English People* (Oxford: Blackwell, 1984/1994).

Schmitt, J-C and J. Baschet, 'La "sexualité" du Christ', *Annales. Economies Sociétés Civilisations*, (2, 1991) 337–46.

Schoeck, R. J. 'Humanism in England', *Renaissance Humanism. Foundations, Forms, and Legacy. Vol. 2. Humanism Beyond Italy*. Ed. Rabil, A. Jr. (Philadelphia: University of Pennsylvania Press, 1988), 5–38.

Schoenfeldt, M. C. *Bodies and Selves in Early Modern England. Physiology and Inwardness in Spenser, Shakespeare, Herbert, and Milton* (Cambridge: Cambridge University Press, 1999).

Screech, M. A. *Laughter at the Foot of the Cross* (Boulder: Westview Press, 1997/1999).

Scribner, R. 'Anticlericalism and the Cities', *Anticlericalism in Late Medieval and Early Modern Europe*. Ed. Dykema, P. and H. A. Oberman (Leiden: E. J. Brill, 1993), 147–66.

Shuger, D. K. *Habits of Thought in the English Renaissance. Religion, Politics, and the Dominant Culture* (Berkeley, Los Angeles, Oxford: University of California Press, 1990).

Siraisi, N. *Medieval & Early Renaissance Medicine. An Introduction to Knowledge and Practice* (Chicago: The University of Chicago Press, 1990).

Smith, L. B. *Henry VIII. The Mask of Royalty* (London: Jonathan Cape, 1971).

Smith, L. B. *Treason in Tudor England. Politics and Paranoia* (London: Jonathan Cape, 1986).

Smith, R. B. *Land and Politics in the England of Henry VIII. The West Riding of Yorkshire: 1530–46* (London: Oxford University Press, 1970).

Sommerville, M. R. *Sex & Subjection. Attitudes to Women in Early-Modern Society* (London, New York, Sydney, Auckland: Arnold, 1995).

Stafford, P. 'Queens, Nunneries and Reforming Churchmen: Gender, Religious Status and Reform in Tenth- and Eleventh-Century England', *Past & Present*, 163 (1999) 3–35.

Stewart, A. *Close Readers. Humanism and Sodomy in Early Modern England* (Princeton: Princeton University Press, 1997).

Swanson, R. N. 'Angels Incarnate: Clergy and Masculinity from Gregorian Reform to Reformation', *Masculinity in Medieval Europe*. Ed. Hadley, D. M. (London and New York: Longman, 1999), 160–77.

Swanson, R. N. *Church and Society in Late Medieval England* (Oxford: Blackwell, 1989).

Swanson, R. N. 'Problems of the Priesthood in Pre-Reformation England', *The English Historical Review*, CV (417, 1990) 845–69.

Swanson, R. N. *Religion and Devotion in Europe c. 1215–c. 1515* (Cambridge: Cambridge University Press, 1995).

Szittya, P. R. *The Antifraternal Tradition in Medieval Literature* (Princeton: Princeton University Press, 1986).

Thomas, K. *Man and the Natural World. Changing Attitudes in England 1500–1800* (Harmondsworth: Penguin, 1984).

Thomas, K. *Religion and the Decline of Magic. Studies in Popular Beliefs in Sixteenth and Seventeenth Century England* (London: Weidenfeld & Nicolson, 1971/1997).

Thompson, B. 'Habendum et Tenendum: Lay and Ecclesiastical Attitudes to the Property of the Church', *Religious Beliefs and Ecclesiastical Careers in Late Medieval England*. Proceedings of the Conference held at Strawberry Hill, Easter 1989. Ed. Harper-Bill, C. (Woodbridge: Boydell, 1991), 197–238.

Thomson, J. A. F. 'Knightly Piety and the Margins of Lollardy', *Lollardy and the Gentry in the Later Middle Ages*. Ed. Aston, M. and C. Richmond (Stroud: Sutton Publishing, 1997), 95–111.

Tougher, S. F. 'Images of Effeminate Men: the Case of Byzantine Eunuchs', *Masculinity in Medieval Europe*. Ed. Hadley, D. M. (London and New York: Longman, 1999), 89–100.

Tudor-Craig, P. 'Henry VIII and King David', *Early Tudor England*. Proceeding of the 1987 Harlaxton Symposium. Ed. Williams, D. (Woodbridge: Boydell, 1989), 183–205.

Underwood, M. G. 'Politics and Piety in the Household of Lady Margaret Beaufort', *The Journal of Ecclesiastical History*, 38 (1987) 39–52.

Vandenbroucke, F. *Le morale monastique du XIe au XVIe siècle* (Louvain & Lille: Editions Nauwelaerts, 1966).

Virtanen, K. *Kulttuurihistoria – tie kokonaisvaltaiseen historiaan* (Turku: Department of History, University of Turku, 1987).

Voci, A. M. *Petrarca e la vita religiosa: il mito umanista della vita eremitica* (Roma: Istituto storico italiano. L'età moderna e contemporanea, 1983).

Walker, G. *John Skelton and the Politics of the 1520s* (Cambridge: Cambridge University Press, 1988).

Walker, G. *Plays of Persuasion. Drama and Politics at the Court of Henry VIII* (Cambridge: Cambridge University Press, 1991).

Walsham, A. *Providence in Early Modern England* (Oxford: Oxford University Press, 1999).

Warner, M. *Joan of Arc. The Image of Female Heroism* (London: Weidenfeld and Nicolson, 1981).

Warnicke, R. M. 'Sexual Heresy at the Court of Henry VIII', *The Journal of Ecclesiastical History*, 30 (1987) 247–68.

Watt, D. *Secretaries of God. Women Prophets in Late Medieval and Early Modern England* (Cambridge: D. S. Brewer, 1997).

White, H. C. *Social Criticism in Popular Religious Literature of the Sixteenth Century* (New York: Macmillan, 1944).

Whiting, R. *Local Responses to the English Reformation* (Basingstoke and London: Macmillan – now Palgrave Macmillan, 1998).

Wiesner, M. E. *Gender, Church, and State in Early Modern Germany* (London and New York: Longman, 1998).

Wiesner, M. E. 'Women and the Reformation in Germany', *Women in Reformation and Counter-Reformation Europe. Public and Private Worlds*. Ed. Marshall, S. (Bloomington and Indianapolis: Indiana University Press, 1989), 8–28.

Willen, D. 'Revisionism Revised', *Religion and the English People 1500–1640. New Voices, New Perspectives*. Ed. Carlson, E. J. (Kirksville, Missouri: Sixteenth Century Essays & Studies, 1998), 287–94.

Willen, D. 'Women and Religion in Early Medieval England', *Women in Reformation and Counter-Reformation Europe. Public and Private Worlds*. Ed. Marshall, S. (Bloomington and Indianapolis: Indiana University Press, 1989), 140–88.

Wogan-Browne, J. 'Chaste Bodies: Frames and Experiences', *Framing Medieval Bodies*. Ed. Kay, S. and M. Rubin (Manchester: Manchester University Press, 1994), 24–42.

Wollasch, J. *Mönchtum des Mittelalters zwischen Kirche und Welt* (München: Wilhelm Fink, 1973).

Wooding, L. E. C. *Rethinking Catholicism in Reformation England* (Oxford: Clarendon Press, 2000).

Wrightson, K. 'The Politics of the Parish in Early Modern England', *The Experience of Authority in Early Modern England*. Ed. Griffiths, P., A. Fox and S. Hindle (Houndmills and London: Macmillan – now Palgrave Macmillan, 1996), 10–46.

Wunder, H. *He Is the Sun, She Is the Moon. Women in Early Modern Germany* (Cambridge, Massachusetts & London: Harvard University Press, 1998).

Youings, J. *The Dissolution of the Monasteries* (London: George Allen and Unwin, 1971).

Index